Environmental Science and Engineering

Environmental Science

Series editors

Ulrich Förstner, Hamburg, Germany
Wim H. Rulkens, Wageningen, The Netherlands
Wim Salomons, Haren, The Netherlands

More information about this series at http://www.springer.com/series/3234

Sairan Bayandinova · Zheken Mamutov
Gulnura Issanova

Man-Made Ecology of East Kazakhstan

 Springer

Sairan Bayandinova
Faculty of Geography and Environmental
   Sciences
Al-Farabi Kazakh National University
Almaty
Kazakhstan

Zheken Mamutov
Faculty of Geography and Environmental
   Sciences
Al-Farabi Kazakh National University
Almaty
Kazakhstan

Gulnura Issanova
Research Center of Ecology
   and Environment of Central Asia
   (Almaty), State Key Laboratory of Desert
   and Oasis Ecology, Xinjiang Institute
   of Ecology and Geography
Chinese Academy of Sciences
Urumqi
China

and

U.U. Uspanov Kazakh Research Institute
   of Soil Science and Agrochemistry
Almaty
Kazakhstan

ISSN 1863-5520   ISSN 1863-5539   (electronic)
Environmental Science and Engineering
ISSN 1431-6250
Environmental Science
ISBN 978-981-13-4861-7   ISBN 978-981-10-6346-6   (eBook)
https://doi.org/10.1007/978-981-10-6346-6

Printed on acid-free paper

This Springer imprint is published by Springer Nature
The registered company is Springer Nature Singapore Pte Ltd.
The registered company address is: 152 Beach Road, #21-01/04 Gateway East, Singapore 189721, Singapore

Preface

This book highlights the studies of differentiation problems of natural geosystems in East Kazakhstan which have an anthropogenic impact. The systematic methodology of comprehensive ecological assessment of anthropogenic impact on natural geosystems and their differentiations on the level of technogenic conditionality for ensuring rational environmental management and environmental protection is stated. The basis for the developments of geosystem approach allows to define stability of geosystems in space and time, which major factor of the organization is nature of lithogenesis and superficial drain interrelations and combination of gravel properties, and technique of geoecological division on the basis of target function creation concerning complex ecological assessments. Despite the abundance of the publications devoted to environmental problems, influence of a technogenesis on the environment is still poorly studied. Therefore, our research results of research can be used by research institutes at a further detailed geoecological research of the geosystems functioning dynamic under the technogenesis influence. Cartographic materials and offered nature protection activities allow developing optimal variants of problem solution on complex use of natural resources, and also can be used by the production, scientific, and other organizations setting the problems solution purpose concerning environmental protection and rational environmental management.

The anthropogenic impact and influence of a technogenesis on the environment and landscape components of East Kazakhstan are considered in the book. The idea of an acceptable environmental risk in the functioning of natural and man-made systems in modern society is given. As well as methodical bases of quantitative assessment of danger in environmental and technogenic risks are given. The traditional methods in geoecology and physical geography were used in this study.

The main purpose of the study is the complex ecological assessment of anthropogenic impact on natural geosystems and their differentiations on the technogenic conditionality level for ensuring rational environmental management and environmental protection.

The practical importance of the book consists of possibility of development of evidence-based recommendations and actions for conservation, quality management of the environment in order to decrease a degree of anthropogenic impact, and prevention of degradation processes.

The book can be useful to the research institutions, industrial, scientific, and other organizations establishing the purpose of the problem solution in environmental protection and rational environmental management. The offered cartographic materials on protection of the nature allow to develop optimal variants of the solution of tasks in comprehensive use of natural resources.

Almaty, Kazakhstan	Sairan Bayandinova
Almaty, Kazakhstan	Zheken Mamutov
Urumqi, China/Almaty, Kazakhstan	Gulnura Issanova

Content and Structure of the Book

The outcomes of studies and research results in this book related to natural geosystems in East Kazakhstan are influenced by anthropogenic impact. The book has four chapters. Chapter 1, "Natural Factors of Forming and Development of Geosystems in East Kazakhstan" considers an overview of natural factors such as geological and tectonic features, topography, climate, soil and vegetation covers, landscape structure in formation of geosystems in East Kazakhstan. The chapter contains eight sections that describe the formation and development of natural geosystems in East Kazakhstan. Chapter 2, "Technogenic Conditionality in Development of Geosystems in East Kazakhstan" provides information and analysis on allocation methods of technogenic geosystems, theoretical substantiation for the organization of geosystems in East Kazakhstan, principles of identification and differentiation of geosystems in East Kazakhstan, characteristic of geosystems, geochemical analysis of technogenic impact and factors of technogenesis. Chapter 3, "Division of the Territory of East Kazakhstan According to the Level of Anthropogenic Impact" contains three sections providing information and analysis on methods in landscape and ecological division based on criterion function, sources of anthropogenic impact on geosystems in East Kazakhstan, division of the territory of East Kazakhstan according to the anthropogenic impact. Chapter 4, "Geoecological Bases of Nature Protection Measures and Actions" has three sections that consider problems and systems of nature protection activities and measures in East Kazakhstan. Chapter 5 contains "Conclusion" and Appendix.

Contents

About the Authors

Sairan Bayandinova is a candidate of geographical sciences and an associate professor.

She studied bachelor and master degrees in Geography at the East Kazakhstan State University. In 2007, she studied a full-time postgraduate program at the Al-Farabi Kazakh National University and has defended a dissertation entitled "Technogenic conditionality of geosystems' development of East Kazakhstan".

Since 2004 she has been working as an associate professor at the Recreational Geography and Tourism Department of Geography and Environmental Management Faculty. She is a director of Student Service Centre «Keremet» in Al-Farabi Kazakh National University.

Her scientific interests are technogenic ecology, geoecology, alternative power engineering, and IT technologies.

She is an author of more than 70 scientific works in domestic, international and rating journals and conferences including four education guidances in Kazakh language such as "Technogenic Ecology" (2012; 2014), "Tourism Industry" (2015), "Logistics in Tourism" (2016).

She has participated in and coordinated the International and Local Projects and Programs such as the International project 543808-TEMPUS-1-2013-1-BE-TEMPUS-JPHES; Professional Training in the area of Information and Communication technologies in Russia and Kazakhstan based on the European standards of qualification (2013–2016); Basic Researches in the area of Natural Sciences (assessment of soil energy); Development of Theoretical Bases and Efficiency Evaluation of use of Innovative Nature Protection Technologies for Recovery of Quality of Natural and Economic Systems in Almaty region (2012–2013); Ecological and Geomorphological Systems of platform-denudation plains in mining regions of Arid Zone of Kazakhstan (2012–2014); Development of mechanisms and guarantees for implementation of investments into forming of the objects of innovative infrastructure providing use of renewable energy resources and energy saving (2012–2014).

Since 2012, she has been the expert of projects by Ministry of Science and Education of Kazakhstan.

In 2004, she became the winner in the nomination "The Best Young Scientist of Al-Farabi Kazakh National University". In 2011, she became the winner of "Talented Young Scientists" by Ministry of Education and Science of Kazakhstan. In 2015, she became the prize winner in the nomination "A leader of sales-2014 for the best educational publication in the Al-Farabi Kazakh National University". In 2016, she became the winner of "The Best Teacher of Higher Education Institution" and awarded the medals "Honourable Educator of the Republic of Kazakhstan".

Zheken Mamutov is a doctor of biological sciences and a professor. In 1962, he graduated from the Abay Kazakh Teacher Training Institute in a major of biology, chemistry, and bases of agricultural production. For many years, he has been working as a teacher at the Abay Kazakh Teacher Training Institute and a leading researcher and scientist at the U.U. Uspanov Kazakh research Institute of Soil Science and Agrochemistry. He has built international cooperation in research and training in young scientists between China and Kazakhstan. He took a position as a Deputy Director and Director at the Green Industry Institute, which was established by his initiative. The main purpose of the Institute was study and development of the saline soils in China. Now he is a permanent scientific consultant of this Institute. Since 2002, he is a professor of the Physical Geography Department at the Al-Farabi Kazakh National University. He has developed the new theoretical basis of melioration of the saline soils; biochemistry of violation of carbonate balance in the saline soils of rice fields. Based on the developed theory, a number of methods of alkalinity regulation in saline soils and irrigating waters of rice fields are offered. The new technology of development of the saline alkaline soils under rice was implemented on more than 100 thousand hectares of rice fields of Kazakhstan, and passed production tests in North Korea, China, Russia, Ukraine, Uzbekistan, and Karakalpakstan. He is an author of a number of methodical developments, which are applied in abroad. He was the first in Kazakhstan who used an electronic microscope, gas and liquid chromatographs, amino-acid analyzers in relation to the studying of saline soils. He has developed a method of identification, a degree of a biochemical capability to alkali formation of irrigating waters in the rice fields. Fourteen candidates and two doctors of sciences have been prepared under his supervision.

His research interests are soil science, ameliorative geography, processes of soil dehumification, adaptive, and landscape system of agriculture.

He is an author of more than 200 publications, including 4 monographs, 15 recommendations for production, and 12 certificates of authorship.

Gulnura Issanova holds a doctorate degree in Natural Sciences and is an associate professor, scientist, and researcher at U.U. Uspanov Kazakh Research Institute of Soil Science and Agrochemistry and a scientific secretary at the Research Centre of Ecology and Environment of Central Asia (Almaty), Kazakhstan.

She studied bachelor's degree (B.Sc.) and master's (M.Sc.) degrees in Physical Geography at the Al-Farabi Kazakh National University and her doctoral degree at Xinjiang Institute of Ecology and Geography, Chinese Academy of Sciences, China.

Her research interest is focused on problems of soil degradation and desertification, in particular, the role of dust and sand storms in the processes of land and soil degradation and desertification. She participates regularly in the International Scientific Activities (Conference,

Forum, and Symposium) on Environmental Problems as well as writes papers on the subject and takes part in local and international projects.

Gulnura Issanova has published many papers in international peer-reviewed journals with high level and wrote a handbook, "How to Write Scientific Papers for International Peer-Reviewed Journals". She is the author of "Aeolian processes as dust storms in the deserts of Central Asia and Kazakhstan" published by Springer Nature, 2017 and co-author of the monograph, "Overview of Central Asian Environments" (in Chinese) and the handbook Methodical Handbook on Interpretation of Saline Soils (in four languages: Kazakh, Russian, English, and Chinese). Gulnura Issanova became a Laureate of the International Award "Springer Top Author" and awarded in the Nomination "Springer Young Scientist Awards" for high publication activity in scientific journals published by Springer Nature, 2016.

Abbreviations

ASRK	Academy of Sciences of the Republic of Kazakhstan
ECMC	Ertis Chemical and Metallurgical Combine
ECSP	Ertis Copper Smelting Plant
EKCCC	East Kazakhstan Copper-Chemical Combine
EKRTDEP	East Kazakhstan Territorial Department of Environmental Protection
MC impacts	Man-caused impacts
MES RK	Ministry of Education and Science of the Republic of Kazakhstan
MPC	Maximum permissible concentrations
MSW	Municipal solid waste
NAPEP	National Action Plan for Environmental Protection
NTC	Natural and territorial complexes
OMPE	Ore mining and processing enterprise
PAH	Polycyclic aromatic hydrocarbons
PB	Polychlorinated biphenyls
RPP	Ridder Polymetallic Plant
SAS	Surface active substances
SNTS	Semipalatinsk nuclear test site
SPNA	Specially protected natural areas
UISP	Ust-Kamenogorsk iron-steel plant
ULZP	Ust-Kamenogorsk lead-zinc plant
UTMP	Ust-Kamenogorsk titanium-magnesium plant

Chapter 1
Natural Factors of Formation and Development of Geosystems in East Kazakhstan

1.1 Background of Research on Geosystems

Kazakhstani part of the Altai, the eastern sides of the Kazakh low hilled part, the basin of Zaisan lake, Tarbagatai, Priertis and other territories related to the administrative division of the East Kazakhstan region are included to territory of East Kazakhstan. Conditionally, we call this territory the East Kazakhstan.

At all times, territory of East Kazakhstan has attracted the attention of many outstanding researchers. It was the area of interest for many geographers: in 1771 P.S. Pallas, in 1829 A. Humboldt, in 1856–1857 P.P. Semenov Tian-Shansky, in 1863–1864 K. Struve and G.N. Potanin, in 1877–1878 N.M. Przhevalsky, in 1903 G.E. Groom-Grzhimailo. It was also visited by the following botanists: in 1734–1741 I. Gmelin, in 1793 I. Sievers, in 1826 K. Ledebour, K. Meyer and A. Bunge, in 1840 G.S. Karelin and I.P. Kirillov, in the years 1895–1911 V.V. Sapozhnikov, 1899–1910 A.N. Sedelnikov, in 1908–1910 and 1936 B.A. Keller; also by following geologists: in 1842 A. Chikhachev, in 1849–1851 A. Vlangali, in 1911–1914 V.A. Obruchev; by zoologists: in 1876 A.E. Brem. Much knowledge about the region was made by an exiled local ethnographer E.M. Michaelis (Kyzykbayev 1964; Chernykh 1971; Klink 1976a, b).

Beginning of knowledge about natural resources of the region started in the 17th century by F.A. Baykov, and in the first quarter of the 18th century by I. Unkovsky. Their travel diaries contained some geographical information about terrain from Tobolsk upwards along Ertis through the mountain ranges of Kalbin and Tarbagatay to Dzhungaria. In the second half of the 18th century, along with expedition of I.G. Gmelin and P.S. Pallas, many researchers such as father and son the Laxmanns, I.M. Renovants, E.M. Patrin, P. Shangin, I. Sievers, F. Ridder, and others worked here. As a result, quite thorough information on orohydrography, soils, flora, fauna, minerals and population was obtained. Most of the materials were connected with Rudny Altai, where mining crafts began to develop in the 18th

© Springer Nature Singapore Pte Ltd. 2018
S. Bayandinova et al., *Man-Made Ecology of East Kazakhstan*,
Environmental Science and Engineering, https://doi.org/10.1007/978-981-10-6346-6_1

century. Routing nature of research and general natural-historical trend are typical for this period (Przhevalsky 1948; Obruchev 1963; Sedelnikov 1898; Klink 1976).

The first comprehensive studies of Altai were carried out by organized expeditions of the Academy of Sciences: as a result of the Gmelin research in 1733–1734 and Pallas in 1771, there appeared description of the river Ertis and Altai structure.

Detailed information on the ores, minerals, general geography and structure of Altai was obtained as a result of Humboldt and Rose trip along Ertis and Altai in 1829. After the 30s of the XIX century, investigation region was exposed to particularly detailed study in connection with ore extraction.

Periods of natural conditions study in the East Kazakhstan can be divided into four parts: (1) Studies conducted during the XVIIIth century; (2) Studies during the most part of the XIXth century; (3) Studies conducted during the late XIX and early XXth centuries before period of agriculture reconstruction; (4) Modern period of research.

The first stage is an epoch of geological observations undertaken by various naturalists, who made their travels with embassy missions in military detachments or in specially equipped expeditions for research.

The second stage is an era of comprehensive study (geography, geology, mineral wealth, botany) of the natural conditions in East Kazakhstan. Studies were conducted both by governmental and public as well as private organizations and companies.

The third stage is the period of industrialization the Greater Altai mastering, systematic study of mineral wealth, hydrogeology and soils by the Geological Committee, organizations of the Ministry of Agriculture and State Property and the Institute of Hydrogeology (Spassky 1809a, b; Beysenova 1982; Tsybulsky 1988a, b; Chikhachev 1974).

Finally, the fourth stage is the systematic era, detailed and comprehensive large scale study of natural conditions by hundreds of researchers, various research institutes and production associations. During this period, methodological foundations of hydrogeology and hydroecology (Sarsenbayev 2001), geo-ecology (Chigarkin 1995, 1996; Kurochina and Makulbekova 1996) were studied in detail.

None of the above mentioned works is specifically devoted to discovery of the nature protection regional aspect. For the first time, these aspects of Kazakhstan geo-ecology were considered by Chigarkin (1974) in his work "The main problems of landscape science and nature conservation in Kazakhstan". In subsequent publications, A.V. Chigarkin has made detailed analysis of principles on geo-ecological zoning and reconstruction, destroyed geosystems, in his works on the East Kazakhstan is considered as geo-ecological province. Regional geo-ecological problems are currently being considered by many scientists of the East-Kazakhstan territorial department of environmental protection (EKTDEP), East Kazakhstan Altai Department of the Geology Institute named after K.I. Satpayev and Al-Farabi Kazakh National University researches. Their works are presented in the project reports.

Thus, after examining the research history on the natural components of geosystems, the following conclusions can be drawn:

- the nature study of the East Kazakhstan from the end of the XVIII century was carried out simultaneously in the form of fragmentary information with reconnaissance works predominance;
- the XIXth century studies are characterized by researchers penetration into almost all naturally historic areas. Also these studies are characterized by comprehensive study of geography, geology, mineral wealth, botany, conducted by both governmental and private organizations and companies;
- studies of the late XIXth and early XXth centuries before the period of agriculture reconstruction differ significantly from the previous two studies by complexity and physical-geographical orientation. Period was characterized by systematic geological study of the East Kazakhstan and its mineral wealth, hydrology and soils conducted by the Geological Committee, organizations of the Ministry of Agriculture and State Property. At that time, main botanical-geographical patterns in distribution of vegetation were established in connection with changes in relief, altitude above the sea level, soil cover nature;
- research undertaken in the modern period, allowed author to objectively present anthropogenesis manifestations on natural system, to use a set of methods that enables to identify the specific manifestations of anthropogenic load.

1.2   Geological and Tectonic Features of Geosystems

Territory of the East Kazakhstan has passed very long and complex path of geological development. In connection with this, it is characterized by structural-tectonic heterogeneity. Proceeding from this, modern hydrogeological processes are determined by the character of Paleozoic and pre-Paleozoic structures and ancient relief and manifestation of young tectonics (Kassin 1941).

Influence of geological structure on the ecological state of the East Kazakhstan is diverse. The main factor is lithology of constituent rocks, their capacity, degree of resistance to man-made impact, porosity and looseness that play a big role in the man-made emission processes of transportation and accumulation (Nekhoroshev 1934), therefore it becomes necessary to consider this section for determining the stability of geosystems for technogenesis (Chigarkin 2003).

The most ancient structures of the East Kazakhstan are Caledonides of the Altai and Shynghyz Tarbagatai zones. Within their boundaries, there are two structural stages: the Blue-Cambrian and the Upper Cambrian-Silurian. Caledonian structures are composed of terrigenous rocks of the Lower Paleozoic to largely Upper Cambrian-Silurian.

The Hercynides of the Ertis-Zaisan geo-synclinal zone extend between the Caledonides of Altai and Shyngyz-Tarbagatai. The Hercynides of the Dzhungar-Balkash geo-synclinal zone adjoin to Shynghyz-Tarbagatai from the southwest.

Within the indicated geo-synclinal zones, two upper structural stages of the Paleozoic are formed: Devonian-Carboniferous and Upper Paleozoic.

Basis for structural plan creation of the East Kazakhstan from very beginning of the geo-synclinal development of this territory was uneven blocky movements of destroyed by tectonic fault of Precambrian base. In direction from the north-east to south-west, following zones of deep faults are distinguished: North-East, Ertis, Shar, Ertis-Zaisan and Balkash-Shyngys (Ayagoz-Urzhar). In addition to enlisted systems of regional deep faults, there is a developed dense network of less significant disturbances and crushing zones, branching off from the main faults and extending parallel to them or in other directions.

Kholzun-Shuy anticlinorium enters the Caledonian system of the mountain Altai and it is separated from the Ertis-Zaisan geosyncline by the North-East zone of crushing. Structure is a complex of linear folds of the north-west strike, made by deposits of Ordovician and Silurian age. Longitudinal faults of the north-west strike are very widespread.

Shyngyz-Tarbagatai meg-anticlinorium is the main structure of the East Kazakhstan, where Cambrian, Ordovician and Silurian deposits are widely developed, crumpled into narrow but tense, sometimes isoclinal folds.

Central strip of the East Kazakhstan is occupied by the Hercynides of the Ertis-Zaisan geo-synclinal zone, within which a number of large structures are distinguished, one of which is the Alei anticlinorium, located in the north-west part of the Rudny Altai. The structure core is composed of metamorphic rocks of Lower Paleozoic, additionally crushed into a series of smaller anticlinal folds of meridional strike.

On described structures of the Caledonian-Hercynian folded base, young structures formed in the Mesozoic stage of geological development on territory are superimposed. Within limits of the Sarsazan synclinorium and Tarbagatai anticlinorium, the Mesozoic troughs partially develop in inherited structures of the Upper Paleozoic structures.

Territory of the East Kazakhstan in the alpine cycle of tecto-genesis developed in isolation on the background of larger geo-structure, which was named Alpine geotectonic system of Altai. Within the latter, three major tectonic structures of the first order are distinguished: the Altai arched uplift (the Altai arch), the Pre-Altai trough and outer chain of the Altai uplifts.

Described tectonic zones of Altai are simultaneous and main orographic systems, which play the watersheds role of the largest Altai rivers.

From the north, north-west, south-west and south, the Altai's arched elevation is bordered by similarly large young negative structure, called the Pre-Delta trough, which has horseshoe-shaped form facing the convex (outer) side to the west. Its foundation on the most of area is composed of hercinides. Following structures of the second order within the Pre-Altai trough are the Kulundy cavity, the Semey-Shar structural jumper and the Zaisan cavity.

The north-east regional tectonic zone is located between the region of the South Altai uplift in the north and area of deep descent of the Paleozoic basement of the Central Zone in south. This regional structure extends in the north-west direction for 250 km and width from 30 to 60 km. Foundation of the North-East zone is composed of intensively dislocated deposits of the lower—middle Carboniferous, interrupted by intrusions of granitoids. Thickness of the young sedimentary cover within described zone varies from 100 to 500 m.

Central tectonic zone represents descended part of the Zaisan cavity, which throughout the entire Cenozoic (and partly both in the Mesozoic and Upper Paleozoic) was stable deflection region. It extends in the north-west direction for 240 km with an average width of 45 km. From the north-east, zone is bounded by the Zaisan regional fault of ancient site, intensively renewed by Alpine tectonic movements (Chigarkin 2003; Nekhoroshev 1936; Bublichenko 1972).

The south-west regional tectonic zone of the Zaisan cavity stretches 180 km with an average width of 40 km in the north-west direction. It is characterized by a shallow uneven lowering of the Paleozoic basement and relatively variable thickness of the Cenozoic sedimentary cover (from 100 to 800 m). In the most descended areas of deflections, there may be deposits of the Upper Paleozoic—the Mesozoic period.

General geological and tectonic features of the East Kazakhstan are basic factors of geosystems spatial distribution allowing to estimate the anthropogenic load. Modern geodynamic features of geosystems also depend on the characteristics of underlying rocks and nature of their occurrence.

1.3   Minerals and Mineral Raw Materials

The East Kazakhstan has large parks of minerals that have direct and indirect ecological impact on the environment: Rudny Altai zone is divided into three metallogenic zones: Beloubinsk-Sarymsakty-Kuritinsk (iron-polymetallic-rare earth elements); Rudny Altai-Ashlyn (gold-copper-polymetallic); Ertis-Fuyunsk (copper-gold ore). The West Kalbin belt is characterized by gold ore, rare earth and rare metal displays. The Zharma-Sauyr belt has three metallogenic zones: Syrektas-Sarsazan-Kobik (copper-gold-rare metal-rare earth); Zharma-Sauyr-Kharatungan (copper-gold ore); Shar-Zimunay (chrome-nickel-mercury-gold ore).

These territories are already or potentially could be objects of technogenic impact with the geosystem disruption. At present, 412 deposits are known, of which 132 are solid minerals and 280 are nonmetallic (Veyts 1959; Shchurovsky 1980).

The Berezov-Belousov ore field combines Berezovsk, Novoberezovsk, Krasnoyarsk, Ertis, Karyerny and Rudny fields, which are mainly composed of volcanic-sedimentary and sedimentary deposits of the Devonian and rarely of early carboniferous.

Polymetallic deposit of Ertis was discovered in 1953, up till 1958 it was explored, and since 1964 it is being worked out underground. The deposit was studied by the discoverers Yu.Yu. Vorobyov and N.I. Stuchevsky. In terms of material composition, the ore is represented by fine and fine-grained differences, composed of intercalations on multicomponent iron, copper, lead and zinc sulphides. The main ore-forming minerals are sphalerite, chalcopyrite, galena, and black ore.

The Belousov deposit is known since 1797, until 1950 its 1 and 2 deposits were worked out. In the years of 1952–1954 the Glubochansky deposits were discovered 3, 4, 5, 6, and in the 70s–7, 9 and 10 (Atlas of the Kazakh SSR 1982; Geologicheskaya izuchennost SSSR 1974) (Tables 1.1 and 1.2). Main ore bodies of the Belousov deposit are predominantly rich and vein-disseminated ores, composed of sphalerite, pyrite, galena, chalcopyrite, and faded ores.

The Nikolaev deposit. The Nikolaev pyrite-copper-zinc deposit in the Priertis district of Rudny Altai is located 12 km south of the Shemonaikhy city. The deposit was discovered in 1749. Study on deposit in different years was carried out by S.G. Ankinovich, N.L. Bublichenko, M.K. Vorontsov, P.F. Ivankin. V.A. Naumov, M.A. Toibazarov (Table 1.3). The main ore minerals are pyrite, sphalerite, chalcopyrite, galena, black ore, nonmetallic-quart, barite, carbonate, etc. (Shcherba 1998).

The Ore deposit. The Orel pyrite-copper-zinc deposit is administratively located near the villages of Gornyak in the Altai territory and Orlovka of the Semey region of the Republic of Kazakhstan. Deposit was discovered in 1959. Deposit was studied by L.I. Pankul, I.T. Sakharov,V.M. Volkov, V.A. Isakov, E.A. Ganja. V.Ivanov, Yu.G. Bondarenko. Leading components of the ores are copper, zinc, lead, sulfur and barium, and the associated components are gold, silver, cadmium, mercury, bismuth, arsenic, antimony, selenium, tellurium and thallium.

The Snegirikhinsky deposit. The Snegirikhinsky pyrite-copper-zinc deposit was discovered in the early 70's. It is located in the same ore field 4 km north-west of the Anisimov deposit source and 14 km south-west of the village Karaguzhyha. The main ore minerals are pyrite, chalcopyrite, galvanite sphalerite and faded ore, secondary and rare are magnetite, arsenopyrite, pyrrhotite, bismuth minerals, etc.

The Chekmar deposit. The Chekmar deposit was discovered in 1976 in the 50 km north of the Ridder city. It is located on the structures continuation of the

Table 1.1 The average content of main components by types and subtypes of the Belousov deposit ores (Geologicheskaya izuchennost SSSR 1974)

Ore subtypes	Basic components,%				
	Copper	Lead	Zinc	Barium	Sulfur total
Polymetallic	0.78	0.73	2.7	2.83	8.21
Lead-zinc	0.36	1.35	4.85	2.97	8.4
Zinc	0.17	0.17	3.10	0.68	6.47
Total polymetallic ores	0.44	0.75	3.55	2.15	7.69
Copper-zinc	1.20	0.18	2.69	1.13	9.8
Copper-pyrite	2.9	0.15	0.17	0.24	18.59
Total for the deposit (balanced ore)	0.54	1.15	4.74	2.65	9.30

Table 1.2 The average content of associated components by types and subtypes of Belousov deposit ores (Geologicheskaya izuchennost SSSR 1974)

Ore subtype	Associated components, g/t							
	Gold	Silver	Cadmium	Arsenic	Antimony	Selenium	Tellurium	Wismut
Polymetallic	0.31	48.40	115.78	593.58	252.38	17.36	4.93	13.90
Lead-zinc	0.72	38.57	169.83	596.57	303.11	9.48	4.35	9.44
Zinc	0.18	9.49	98.70	406.5	104.69	4.93	4.0	6.69
Polymetallic	0.49	45.59	128.10	532.25	220.06	10.59	4.43	10.01
Copper-zinc	0.31	97.96	95.69	684.80	167.12	15.31	4.95	14.13
Copper-pyrite	0.51	8.49	8.0	174.0	186.0	19.1	8.1	17.6
Total	0.38	51.30	158.33	651.44	285.53	14.1	4.54	12.23

Guslyakov deposit in south-east direction in center of the Star-Guslyakov ore zone. Composition within the counting contours is dominated by polymetallic, lead—and copper—zinc differences.

The Ridder ore deposit. Deposits of the Ridder ore field (Sokolnyi, Ridder, Kryukovsk, Ilyinsk, Filippov, and Novoleninogorsk) are located in the east part of the Rudny Altai at the junction of two regional structures—the Caledonides of the Altai-Sayan folded zone and the hercynids of the South-West Altai.

The Ridder-Sokol deposit. The Ridder deposit is known since 1784. The Sokol deposit was discovered in the first half of the 19th century. The deposit is located in the middle most elevated part of the Ridder graben. According to material composition, the ores belong to proper polymetallic type. Copper-pyrite, copper-zinc, polymetallic and barite polymetallic ores are distinguished in ore deposits. The main industrial components are lead, zinc, copper, and associated gold, silver, cadmium and number of other impurity elements.

The Novoleninogorsk deposit. The Novoleninogorsk deposit is located in the east part of the Ridder ore deposit, in the area of its combination with the Uspensko-Karelinsky zone (13 km east of the Ridder city). It was discovered in 1981 and studied by G.S. Janvarev, A.M. Kudryavtsev. The Novoleninogorsk ore deposits are mainly vein-disseminated, polymetallic, lead-zinc with gold and silver.

The Tishinsky deposit. The deposit is located 18 km south-west of Ridder, in the central part of the Butachikha-Kedrov zone. It is located on the south-west wing of the Sinyusha anticlinorium and is an integral part of the Rudny Altai polymetallic belt. The Tishinsky ore deposit is confined to the zones intersection of sublatitudinal faults and the Butachikhinsky-Kedrov deep fault of the north-west strike. Ore deposits are pyrite-polymetallic.

The Maleev deposit. The Maleev pyrite-polymetallic deposit is located in 12 km to the north of Zyryanovsk. It was known since 1840 and on small scale was reconnoitered and worked out until 1954. Resumption of geological exploration in the 1980s made it possible to discover the ore-bearing zone and to attribute deposit to category of large ones.

The Zyryanov deposit. It is located in the same area of Rudny Altai, on west slope of the Revnyushinsky structure. The deposit was discovered in 1791, studied

Table 1.3 Characteristics of Nikolaev deposit ore types (Geologicheskaya izuchennost SSSR 1974)

Ore types	Minerals		Characteristic	
	Main	Secondary	Texture	Structures
Crystalline	Pyrite Chalcopyrite Sphalerite Quartz Sericite	Marcasite Galena Barite Sericite The faded ore	Massive, nested-interspersed, veined crushing	Hypidiomorph-nonallotrio-morfno-granular, emulsion
Transitional	Pyrite Marcasite Sphalerite Wurzit Chalcopyrite	Galena Sericite The faded ore Quartz	Breccia, cementation, patchy, nest mesh	Breccia, cementation, patchy, nest mesh
Meta Colloidal	Marcasite Wurzit Melnikovit Chalcopyrite Galena Thin-grained pyrite with marcasite	Sphalerite The faded ore Calcite Cowellin Chalcosine Greenokite Barite	Lenticular-lamellar, clots, rhythmically-layered, colloform-banded, reniform, brecciated	Colloid, mossy, rhomboidal, finely dispersed radial-radiant, granular

by many generations of geologists. More than 100 minerals are known in the ores. Main ore minerals are sphalerite, galena, chalcopyrite and pyrite, nonmetallic—quartz, calcite, dolomite, barite, etc. Main useful components in the ores are copper, lead and zinc, content of which varies from fractions to tens of percent.

The Karchig deposit. The Karchig deposit is spatially located in axial part of the Kurchum-Kaldzhir horst-anticlinorium. The deposit was studied in 1913–1914, later with interruptions in 1934–1969. In the study and deposit evaluation, participation was attended by G.G. Kell, D.M. Shilin, B.I. Veits, A.N. Derbas, A.A. Shatobin, B.F. Zlenko. In the ores composition, as in the Babylonian deposit, a large number of minerals are noted. Main ore minerals are pyrite, pyrrhotite, chalcopyrite, sphalerite, magnetite.

The Babylonian deposit. The Babylonian copper-pyrite deposit was developed in ancient times, and later—with interruptions in 1786–1817. Since 1940, it has been studied by A.P. Nikolsky, N.N. Velikaya, B.I. Veits and other researchers. According to these researchers, the alternation of quartz-cordierite-anthophyllite, actinolite and quartz-micaceous schists, graphitized phyllites, and quartz-feldspar biotitized sandstones of the Takyr Formation is characteristic of the deposit.

Thus, the analysis of deposits described above, allows us to conclude that territory of the East Kazakhstan with constituent rocks of the pre-Palaeozoic and Paleozoic age belongs to ecologically stable category of (Chigarkin 2003), but at the same time they are objects of intensive technogenic impact with observable disruption of geosystem structure.

1.4 Topography and Modern Geomorphological Processes

Consideration of the orography features will allow characterizing the environmental properties of environment, from position of the most complete anthropogenic impact evaluation.

Surface structure of the East Kazakhstan is divided into six regions. The north-east part of the territory under consideration covers the south-west periphery of the Sayano-Altai mountain system and is known in geological literature under the name South-West Altai. In the south-east, there stretches the plain of the Zaisan intermontane depression, framed from the south and south-west by ridges of Sauyr-Tarbagatai. In extreme south of the East Kazakhstan, there is a desert-steppe plain of the Alakol intermontane depression. The west and south-west parts of described territory are occupied by extensive elevated small-sap plains of the Central Kazakhstan. Gradually decreasing in the north and north-west directions, the Centra Kazakhstan low hill terrain in the Priertis Semey region merges with vast plain of the West Siberian lowland.

The South-West Altai is a mountainous country. On the north border, there are ridges of the so-called Mountain Altai, which serve as a watershed of the rivers Ob and Ertis. The highest of them is the Katun squirrel ridge with the main peak Belukha Mountain (4605 m). From this orographic center, in the north-west direction extends continuous chain of ridges: Listvyaga, Holzun, Koksu mountains and Tigiretsky range. Their heights are 2200–2800 m and decrease to the south-west. Main watershed from the south-west adjoins a complex system of mountain ranges and ridges, among which the most significant ranges are Ulba (1800–2200 m), Ivanov (up to 2800 m) and Uba (1100–2000 m), having southwest strike and forming watersheds of right-bank tributaries of Ertis—Uba, Ulba and Bukhtyrma. These ridges for the wealth of their subsurface were named Rudny Altai.

To the south of Altai mountains, there is a system of the South Altai ridges; here, in the direction from north to south, the Tarbagatai ranges (2200–3000 m), Sarymsakty (3000–3400 m), Naryn (1500–2500 m), South Altai (2800–3400 m), Kurshym (up to 2000 m) and Azutau (2000–2300 m).

In the south-west of region, there is the Kalbin ridge (1600 m), elongated in latitudinal direction with turn to the west.

Within the South-West Altai, high, middle and low-mountain types of mountain relief are encountered. The high-mountainous relief, developed within the boundaries of South Altai at absolute elevations above 2000 m, is characterized by extremely dense dissected and large depth of the cut (up to 1700–1800 m), mostly crest-shaped, less often set peaks, steep naked slopes. Within the ridge, present-day glaciation has been preserved. Here according to Tronov (1949), there are 122 glaciers. Relics of glaciation in the form of typical glacial relief forms are found within mountainous and South Altai.

The South Altai is characterized by extreme steepness of slopes, some leveling of the watersheds in far west and strong dissection of them with abundance of glacial forms in the central and east parts. Relative elevations vary within 600–1000 m, reaching in some cases 1500–1700 m. The highest elevated parts of the Naryn, Sarymsakty and South Altai ranges are complicated by Alpine relief forms.

Mid-mountain relief, widely developed in the South-West Altai, is characterized by absolute elevations of 1000–2000 m and differs from high-mountain relief by shallower depth of erosion cut (maximum values here do not exceed 500–600 m), somewhat less than the subdivision degree. Distinctive feature of the mid-mountain relief is the appearance in lower parts of mountain slopes, in logs and valleys of rivers, clusters of loose detrital, poorly sorted deluvial-proluvial and alluvial-proluvial material up to several meters in thickness.

Low-mountainous relief is the most widely developed within described region and represents transition stage from the described high-mountain and mid-mountain relief types to the foothills hillocky. Absolute marks of watersheds and individual peaks fluctuate within 600–1000 m. Relief of the most low mountains is slightly dissected, flattened, with flattened watersheds and peaks of separate hills. The maximum depth of cut is 100–200 m.

Within the low mountains, mature forms of river valleys are observed by well-developed channels and complex of accumulative terraces, thickness of deposits, which reaches tens and meters.

Large rivers (Ertis and Bukhtyrma) have up to six terraces above the floodplain from the Lower Quaternary to the present, and two levels of floodplain terraces, with high levels of terraces remaining only fragmentarily in sections of several hundred meters in length. The height of terraces increases from 2 to 5 m in low terraces above the floodplain to 60–80 m at the high. The width of terraces ranges from hundreds of meters (high terraces) to several kilometers (low terraces).

One of the characteristic features of the mountainous part relief of the East Kazakhstan is presence in it the number of intermontane depressions (Ridder, Naryn-Bukhtyrma, Verkhnyi-Karakabin, Bobrov, Uspen, and etc.), due to their origin of tectonics and filled with loose Cenozoic deposits. The relief of depressions is generally flat. Surface of the basins gradually sinks from the east to west: Ridder from 1000 to 600 m, Naryn-Bukhtyrma from 1000–1300 to 800 m, Verkhnyi-Karakabin from 2300–2600 to 1600 m; surface of the Bobrov depression is inclined from 1500 to 1100 m from the west to east and from the north to south.

The Zaisan depression is located between the intermountain constructions of Altai and Kalba—in the north, and Sauyr-Tarbagatai—in the south. Relief of the most territory of the Zaisan depression has the appearance of accumulative alluvial-proluvial and alluvial-lacustrine plain. The flat surface of plain is slightly inclined to depression center, where the lake is located. Zaisan, now closed by the Bukhtyrma reservoir, with an absolute mark of the water mirror of 394.8 m. Erosion-denudation-ridge-hilly plains are widespread in the central part of Northern Prizaisan and on the left bank of Kara Ertis.

Restricting from the south of Zaisan depression, system of ridges that are part of the Sauyr-Tarbagatai, stretched in latitudinal direction, is represented by high-mountain, middle-mountain and low-mountain relief types. In the east, mountain system begins with the Sauyr ridge. The Sauyr is the highest ridge in described mountain system (up to 3816 m), with adjacent region of modern glaciation of Sauyr-Tarbagatai. Position of snow line is about 3300 m; boundaries of glacial tongues are about 3000–3500 m. Total area of glacial, firn and snow fields are about 30 km². At altitudes of 3200–3700 m, the forms of dissected glacial-alpine relief are developed.

From the Sauyr Ridge, there branches out the lower Manyrak Ridge in the west and north-west direction, which has maximum height of 2053 m. The east branch of the Manyrak has contours of massive blocky uplift, known as the Saikan Ridge. These ridges are actually the Sauyr Mountain Group.

The Tarbagatai ridge stretches in the west of region, its axis is experiencing general dip in the west direction, due to which absolute marks of watershed decrease from 2900 to 1600 m from the east to west. In the west and north-west direction, it passes into low relief with soft contours, close to hillocky type.

Between the ridges of Saikan and Sauyr there is the Kenderlyk trough, in the central part of which there extends the Akzhal ridge. Along axis, in the north-west direction, it is stretched for 25 km and reaches width in the middle part of 14 km. The absolute marks of surface are 1000–1600 m, and the Akzhal ridge—from 2000 m in the east to 1300–1400 m in the west. To the south, hillocks gradually and imperceptibly transform to the Shyngyz hills, acquiring some elements of low mountains relief. Here, in the north-west direction, the Khan-Shyngyz and Shyngyz-Tau ridges stretch north-westward, coinciding with direction of the

Sauyr-Tarbagatai ridges. These mountains are main watershed between the river basins Ertis and the Balkash lake.

There are several types of relief among the upland raised denudation plains. The most widespread is tablelike steepy sloping relief of various hypsometric levels; surface profile of ridges is symmetrical with flat-lying slopes (5°–10°), with relative elevations up to 100 m.

The West Siberian lowland enters borders of the East Kazakhstan by extreme south-east part and is monotonous flat forest-steppe and steppe plain of the Ertis Semey region, in the south part with terraces of the river Ertis. In the south-west direction, markings are reduced to 150 m and surface becomes hollowed terraced lacustrine-alluvial and alluvial plain.

The above-described features of surface structure of the East Kazakhstan, and modern geomorphological processes, play a significant role in the structural organization of geosystems confined to the river basin Ertis and its tributaries of various orders. Pattern of ancient surface drain often does not correspond to modern hydrographic network, but plays a leading role in nature of geosystems structure and the degree of functioning intensity, which is associated with processes of their self-regulation.

Thus, characterizing the ecological consequences of orography and relief, one can come to conclusion that the middle flow of the river Ertis is a transit zone for the pollutants removal; location of the mountains determines the barrier climatic effect, occurred in accumulation on windward slopes of atmospheric precipitation, which results in the drain formation of mountains leading to exogenous occurrences, mudflow and other processes of natural environmental ecological destabilization; the presence of denudation and accumulative plains contribute to broad development of water erosion processes (Nekhoroshev 1934; Shcherba et al. 2000; Bublichenko 1936; Obruchev 1927; Kassin 1947; Obschaya harakteristika geomorfologii 1947).

1.5 Climate Conditions in Formation of Geosystems

Climate of the East Kazakhstan is continental with large daily, seasonal and annual amplitudes of air temperature fluctuations, which is determined by deep inland location of territory.

In warm season the radiation balance is everywhere, except the alpine belt, and it is positive with the largest values in June and July. Negative balance is observed everywhere from November to February and reaches the lowest values in December and January. The maximum monthly averages of radiation balance can reach from 9.8 to 10 J/m^2, and the minimum from 1.7 to 1.9 J/m^2. Total annual values of the radiation balance vary from 37 J/m^2, in highlands to 42 J/m^2, in the northern foothill areas, and reach 40 J/m^2 in the south of the Zaisan depression, in the Ertis and Bukhtyrma valleys up to 43 J/m^2 (Kaletskaya et al. 1945).

Total solar radiation across the territory varies naturally from north to south and is in the north 108–110, in the river valley Bukhtyrma—122–124, and in the south —134–135 kcal/cm^2 per year.

By the sky clarity and the sunshine hour's number, the East subregion surpasses the European part of the Commonwealth countries territory at the same latitudes and can compete with the Crimea. For example, the cities of Kiev, Ust-Kamenogorsk and Ridder are located in the same latitude. The annual number of hours of sunshine is 1786 in Kiev, 2287—in Ust-Kamenogorsk, 2395—in Ridder.

East Kazakhstan can be attributed generally to the well-hydrated areas of Kazakhstan, if you consider that 30% of its territory receives less than 200 mm of rainfall a year and only 20% of the territory—400 mm/year (Uteshev 1959).

Annual amounts of atmospheric precipitation, over the subregion territory, vary from 119 to 220 mm in central parts of the Zaisan depression. On foothill plains of the Rudny Altai, they vary from 300 mm to 500 mm, and height in the mountains reach 2000–2500 mm. In the South Altai, they may fall to 1200–1500 mm. So, the mountains become an obstacle or kind of barrier in the air masses way. Such barrier is formed, when the air masses are forced to rise along windward mountains slopes.

Monthly maximum precipitation is most often observed in June or July. In the South-West Altai, there is a second maximum, and it is less expressed, in October and November. In the remaining areas of the Mountainous Altai and Upper Ertis from summer to autumn, there is a gradual decrease in the precipitation amount (Tables 1.4 and 1.5).

The greatest amount of precipitation falls in the area of the river Malaya Ulba, where annual precipitation is 1500–2000 mm and more (according to the observations of hydrometeorological station, in 1979 their amount was 4000 mm). The second region of large precipitation is located in the South Altai, it is also the glaciation center. From the station, the greatest amount of precipitation falls in Ridder (675 mm), in Zyryanovsk (605 mm) and Ust-Kamenogorsk (498 mm).

Mid-mountainous and high-mountainous areas of the Rudny and South Altai, the central part of Kalba and Sauyr belong to the sufficient moisture zone. In mountainous areas, the amount of precipitation reaches 1000 mm and even 1500 mm a year. It is these precipitations that feed the rivers that originate in Altai.

In the direction from mountains to the Ertis valley, amount of precipitation decreases sharply. In the middle flow of the Bukhtyrma, they are only 400 mm. And within the east, higher part of Kalba, rainfalls reach 700–800 mm, and in the foothills they decrease to 300–400 mm. The lowest amount of precipitation is recorded in Zaisan (311 mm) (Egorina 2004; Scientific and application-oriented reference manual on climate of the USSR 1989).

Significant amount of precipitation falls in the snow form (Table 1.6). The average height of snow covers during the cold season, and reaches 86 cm in Zyryanovsk, also remaining high (about 0.5 m) at other stations. In general, it should be noted that the East Kazakhstan is characterized by presence of stable snow cover for 5 months per year and more (Table 1.6).

Table 1.4 Monthly, annual and seasonal amounts of precipitation, mm

Station	Height, м	Monthly amounts of precipitation												(XI–III)	(IV–X)	Year
		I	II	III	IV	V	VI	VI	VIII	IX	X	XI	XI			
Upper Ertis																
Semey	195	18	14	17	18	24	35	37	25	19	25	28	23	100	183	283
Kokpekty	510	24	19	18	18	26	31	33	27	18	22	35	35	131	175	306
Bukhtyrma	373	19	20	28	30	41	54	57	41	28	41	42	33	142	292	434
Katon-Karagai	1081	16	12	14	26	56	63	67	58	36	35	27	22	91	341	432
Buran	409	10	9	12	16	20	18	19	15	13	20	21	16	68	121	189
Zaisan	604	9	9	16	28	40	41	37	27	23	24	21	16	71	220	291
Abai	617	11	10	13	20	28	38	42	23	13	11	20	16	70	175	245
Zharma	678	13	12	18	23	31	31	43	30	21	23	28	21	92	202	294

Table 1.5   Monthly and annual rainfall (mm)

Station	Months												Year
	1	2	3	4	5	6	7	8	9	10	11	12	
Ust-Kamenogorsk	25	26	33	35	48	56	62	49	36	46	46	36	498
Ridder	17	17	25	52	87	87	10	82	70	67	43	24	675
Zyryanovsk	42	33	31	38	60	60	72	54	43	60	58	54	605
Shemonaikhy	24	25	27	33	47	49	60	42	33	44	41	35	460
Samarka	26	22	23	26	35	41	46	34	29	38	47	33	400
Zaisan	10	9	17	32	42	42	40	29	25	27	21	17	311
Semey	19	16	20	18	26	37	40	28	20	28	30	24	306

Table 1.6   Snow cover height (cm)

Station	Months							Per year		
	10	11	12	1	2	3	4	Avg	Max	Min
Ust-Kamenogorsk	1	12	28	40	5	32	13	57	93	11
Ridder	4	17	27	33	3	29	17	44	87	7
Zyryanovsk	1	24	51	67	7	70	43	86	132	47
Shemonaikhy	1	12	30	37	4	27	11	47	98	10
Samarka	1	16	32	44	5	35	13	57	98	23
Zaisan		7	15	18	1	10	3	22	42	6
Semey		6	15	19	2	13		27	83	5

It is known that orographically difficult regions are significant obstacle to air flows. The average wind speed per year does not exceed 3 m/c at any of the stations. At Zyryanovsk station it is only 0.8 m/c.

Here the influence of orography is clearly seen: wind blows along the valley of Ertis up or down the valley. Repeatability of windless conditions is somewhat lower than in Zaisan, but high—31% per year. In winter it exceeds 40%. In the mountains of the East Kazakhstan, at elevations from 600 to 1400 m, belt of increased clear sky frequency and intense solar radiation is allocated.

Continentality of the subregion climate is emphasized by large amplitudes of annual and diurnal temperatures. The annual temperature amplitude in the Orlov settlement is 43 °C, in Zyryanovsk 42 °C, on the foothill plains is 35–37 °C, on the slopes of the mountains—about 30 °C.

The average annual air temperature ranges from 3.0 °C to −3.6 °C in the lowland south-west regions and near the large reservoirs (Zaisan), to −6 to 7 °C in the high-mountain areas (Table 1.7). In plains and foothill-lowland areas, the average annual air temperature is increasing as it moves from the north to south and from the east to west.

The average temperature of the warmest month (July) overall (except for the highlands) exceeds 15 °C, reaching 20–22 °C in dry steppes and semi-deserts in the

Table 1.7 Average monthly and annual air temperature, °C

Station	Height, м	I	II	III	IV	V	YI	VII	VIII	IX	X	XI	XII	Year
Semey	195	−17.1	−16.6	−9.3	3.8	13.0	19.0	20.9	18.6	11.9	3.8	−6.8	−14.1	2
Kokpekty	510	−20.9	−19.5	−11.0	2.6	12.1	17.9	20.4	18.6	11.9	2.4	−9.3	−18.4	0
Bukhtyrma	373	−18.3	−16.0	−9.6	3.3	12.0	17.7	20.4	18.4	12.2	4.4	−7.6	−16.0	1
Katon-Karagai	1081	−14.8	−12.5	−6.1	3.2	10.2	15.1	17.2	15.2	10.6	2.8	−8.5	−13.6	1
Buran	409	−18.8	−16.5	−7.3	6.2	14.4	20.1	22.2	20.0	13.6	4.7	−6.2	−15.0	3
Zaisan	604	−17.8	−15.8	−7.7	5.8	14.2	20.4	22.7	21.4	15.2	5.7	−6.1	−15.0	3
Abai	617	−14.1	−13.8	72	4.1	12.2	17.9	20.3	18.3	12.2	3.7	−6.2	−12.6	2
Zharma	678	−15.4	−15.2	−8.8	3.1	11.4	17.1	19.3	17.3	11.2	3.1	−7.4	−14.0	1

south-west and west of the territory (Table 1.6), in the foothill plains 18–23 °C, on the slopes of the mountains 16–18 °C, in the mountain hollows 14–16 °C. By noon, the air temperature can usually reach 24–26 °C. Absolute maximums are 40–42 °C.

Temperature gradient in July for mountainous areas is 0.5–0.7 0 per 100 m. Near glaciers, the average July temperature does not exceed 6–10 °C. The coldest month is January. Distribution of January temperatures depends on macrocirculation factors and relief. The average temperature in January ranges from −14 to −19 °C. The coldest place in winter on region territory is in the closed Orlov hollow, in the Orlov settlement (Kurshym district), where the average January temperature is −27°, and the absolute minimum is −62 °C.

For the basin of the Upper Ertis, in connection with its position almost in the continent center, very large amplitudes of temperature fluctuations are characteristic. The absolute minimum reaches −62 °C in highlands, and −53 °C in plains. The absolute maximum reaches 42 °C in flat part (Semey), and in highlands it is much smaller (Uteshev 1959).

Vegetation period (with an average daily temperature above 50) continues from the second to the third decade of April and the second decade of May to the end of September or to the end of October. With the height of terrain, duration of the vegetation season decreases from 190 days in foothills to 60 days a year in the highlands.

The first snowfalls and unstable snow cover in the north-west regions are observed in October, in the north-east foothill regions—in September, and in the high-alpine regions of Altai in late August and early September. Steady snow cover is formed on average 20–30 days later.

Occurrence duration of stable snow cover varies from 135 to 150 days in plains and low mountainous regions up to 170 days in the north-east foothills of Altai.

Thus, because of the great distance and fenced off by mountain systems, warm and moist air masses from the Atlantic Ocean reach this area loosing most of their moisture, and air masses entering from the Arctic Ocean reach cold and dry. The hilly, shallow and flat areas of the Ertis river left bank are particularly dry. Complex orography and presence of closed plains and plateaus cause great differences in climatic characteristics of individual regions. Altai's climate is generally much milder than in neighboring areas, summer is cool, and winter is relatively warm.

Climatic conditioning of environmental conditions allows us to draw the following conclusions: the climate of the East Kazakhstan has huge impact on the environment state, major environmental factor that has strong impact on surface waters and groundwater, soil, vegetation and natural-anthropogenic landscapes (the greatest importance is atmospheric air pollution under influence of various natural and anthropogenic factors), peculiarities of air masses circulation on anticyclonic regime favor the atmosphere pollution and aggravate the air basin pollution of large cities and industrial centers of the East Kazakhstan (Ust-Kamenogorsk, Ridder, Zyryanovsk) (Nekhoroshev 1934).

1.6   Analysis of a Ground and Underground Runoff

Study of surface waters of the East Kazakhstan began in 1870, when former Ministry of Railways on the Ertis conducted expeditionary studies to find out shipping conditions. The first water state gauging stations were also opened, which initiated stationary regime observations in the river Ertis (Vodnyie resursyi Kazahstana 1957).

Hydrographic structure of river network in the East Kazakhstan is caused by complex relief and various climatic conditions.

The thickest hydrographic network is characteristic for well-moistened peripheral regions (West and North Altai). In the Ertis river basin, the average density of river network is 0.27 km/km², in some areas of right bank the dissection of territory by riverbeds reaches 0.70–0.75 km/km² (the Uba river basin).

Main rivers of territory under consideration flow in the north and north-west directions, correspondingly to general decrease in terrain, strike of the largest tectonic faults and main watersheds position. River valleys in the east of region are wide, with sloping inclined slopes imperceptibly merging on watersheds with the surrounding terrain.

There are more than 32,650 rivers on territory under consideration, of which one river (Ertis) is more than 1000 km long and four rivers have length of 300–500 km. Remaining rivers are less than 300 km long, with more than half of the total river numbers having a length of less than 10 km (Vodnyie resursyi Kazahstana 1957; Akhmedsafin 1971).

Rivers of Altai are characterized by large flow velocities (from 2 to 5 m/c). Plain rivers have much lower maximum speed—0.8–1.5 m/c. Depths of the rivers are usually shallow. The maximum depths of small and average rivers are 1–2 m and large rivers are 3–10 m.

The river Ertis originates in China. Its sources lie on the west slopes of the Mongolian Altai at the height of 2500 m, and there it is called Kara Ertis (Black Ertis). Kara Ertis flows into the lake Zaisan, and after flowing out of it, river receives the name Ertis. Watershed area of the river within borders of considered territory is 196,000 km². Within the area, its length reaches 1500 km and its width reaches 120–150 m. The river drain is regulated by dams of three hydroelectric power stations, which are Ust-Kamenogorsk, Bukhtyrma and Shulbinsk. The average water consumption at the riverhead of Ertis is 350–370 m³/s, and it's more than 600 m³/s below the Bukhtyrma mouth.

All the large tributaries of Ertis are right-bank, their sources lie on the Altai Ranges. Mostly they are water-abundant, with large gradients and rapid course; there are many rifts in the channels. The largest and most watery tributaries are Bukhtyrma, Uba, Ulba, Hamir, Berel, and others.

On the upper section, before it disembogues itself into the lake Zaisan, river takes a number of tributaries like the river Kurshym and watery Caljir, which flows out of the lake Markakol. Below estuary of the river Bukhtyrma, where Ertis enters narrow mountain canyon, powerful hydroelectric power station dam was erected in

1960. Bukhtyrma reservoir, which is the largest in country and which was formed as a result of the dam support, absorbed the lake Zaisan. Number of large tributaries like Kurshym, Naryn, Buktyrma, Bolshoi Bucon, Tainty flow into the Bukhtyrma reservoir.

Below the second dam, Ust-Kamenogorsk hydroelectric power station, at the outlet of the mountains river, two more right-bank tributaries like the rivers Ulba and Uba flow into it. Length of the river Kurshym is 218 km.

The largest tributaries of Ertis are Bukhtyrma and Uba. Bukhtyrma catchment area is 12 660 km^2, the length is 336 km. Bukhtyrma originates from small lake in the South Altai, and before entering the Bukhtyrma reservoir it accepts a number of tributaries like the rivers Belaya Berel, Belaya, Chernovaya, Cheremoshka, Sarymsakty, Levaya Berezovka.

The river Uba is formed from the confluence of rivers Chernaya Uba and Belaya Uba, originating from the snow-covered Korgon mountain peaks. Catchment area of Uba is 9850 km^2, the length is 278 km. Its largest tributaries are Chesnokovka, Stanovaya, Beloporozhniaya Uba, Malaya Ubinka, and Shemonaikhy. Hydrographic survey in the Ertis basin was carried out on 47 rivers, including 13 rivers in the Uba river basin.

Ulba is formed at confluence of the rivers Filippovka, Bystrukha and Gromotukha. The river flows through wide valley, overgrown with pine and poplar woodlands, falling into the Ertis. Its largest tributary is Malaya Ulba.

The Ertis river left bank originate on the slopes of the Kalba Range. The Kalba hydropower network is relatively poor; the rivers are shallow and shoal in the second half of summer. They flow among stony mountains, covered with mixed forest. Floodplains of rivers are very beautiful, covered with shrub vegetation, poplar, aspen and birch.

In north part of Kalba, small rivers Kyzylsu, Kapan, Ulanka and others carry their waters to the Ertis through bald peaks.

In the south of Kalba, Bolshoi Bukon and Kokpekty are striving to the lake Zaisan. The river Shar flows in the southwest. It has a small reservoir.

There are 1003 lakes in the East Kazakhstan, their size is one hectare or more. Large lakes are 18. They are located unevenly among territory, most of them are in the north part, large ones are in valleys and intermountain basins. There are many small lakes on the mountains. They are shallow, depth varies from 2 to 20 m, but some of them are 30 m deep and more. Lake basins are very diverse by origin. The most typical characteristics for them is that they are tectonic, pond and lateral. Colour of the water varies from bluish-green to yellowish-brown. Water temperature reaches 7–15 °C in the end of May. Highest temperature is 20 °C, it takes place in July and August.

The largest lake in region was the lake Zaisan. But after construction of the Bukhtyrma hydroelectric power station on the Ertis, the outlines of its shores changed dramatically. Waters of Bukhtyrma reservoir were flooded with vast expanses, nature of the lake changed. A lot of names were given to the lake Zaisan. More than 280 years ago it was called Kyzyl-Pu, later it was Kortsan, then it was the sea of Teniz, for the special noise of surf waves it was named as Hoshtu-Nor (Lake of bells), Nor-Zaisan or simply Zaisan. There were other names too.

All lakes of ponds are shallow. There are five of them; they are Kashkerbai (Sasykkol), Alka (Kurzhinkol), Ulmeis (Shalkar), Dyuisen (Tortkarakol), Istykpa (Kanakol). Lakes are located step-wise, at the south edge of the Koktau massif. Basins of them are closed on three sides and, as it were, cut into granites. Total area of the Sebinsk lakes is 5.56 km, depths ranges from 2.5 to 38 m, height above sea level is 770–830 m. Water colour is yellowish or brownish-yellowish. There is an unnamed stream, which flows through all the lakes. They are fed by groundwater, due to river flow and atmospheric precipitation.

Swamps in the considered territory are not so widespread. There are two types of swamps: upper marshes of mountain river's flat watersheds (the Western, Northern, Central, South-Eastern Altai) and wetlands in the floodplains and estuaries of rivers. Wetland floodplains of rivers are found on sections of the river valleys and river crossings of intermontane basins (the Abai steppe marshes). Wet river mouths are most widespread in the Zaisan depression, where they alternate with salt marshs.

Modern glaciations take place mainly in the highland part of Altai. Nowadays, there are 1163 glaciers with total area of about 800 km^2, and more than 250 glaciers discovered on the basis of aerial photographs in the last 20 years. On the range Sauyr, there are 32 glaciers with total area of about 30 km^2. In other mountain ranges (Shyngiz-Tau, Tarbagatai), there are no glaciers because of their comparatively low altitude. Great merit belongs to the scientists of Tomsk University in the study of Altai glaciers. In the Republic of Kazakhstan, glaciers were investigated by employees of the geography sector of the NAS of the RK, K.G. Makarevich, E.N. Vilesov and A.V. Khonin (Sarsenbayev 1971; Katalog lednikov SSSR 1969; Vilesov 1977).

Snow line undergoes fluctuations as it moves from places with high humidity to places with less humidity, and also depending on the exposure of slopes and terrain latitude. More or less constant altitude position determines the firn line dividing glacier and firn basin. Heights of firn and snow lines increase from the north-west to south-east. According to M.V. Tronov, the height of snow line in west ranges of Altai is at an altitude of 2300–2500 m, on Belukha at an altitude of 2750 m, in the extreme east of the Altai at an altitude of 3400–3500 m.

Underground waters are widespread in the territory under consideration and characterized by variety of types, according to the nature of water bearing rocks, circulation and mineralization.

Hydrogeological conditions of the Altai mountains are very diverse. According to the nature of water bearing rocks and the circulation conditions in the Altai, fissured waters of the weathering crust, fissured veins of bedrock, fissure-karst waters, and underground waters of loose sediments are distinguished. Fractured waters of the rock weathering zone and fissure-vein waters are most widespread. The rates of sources from effusive-sedimentary deposits are predominantly equal to 1.0–1.5 l/s. Fissured waters of sandy-shale deposits, widely spread in the north-west and south parts of the Altai, are also comparatively unremarkable. For most sources, the production rate does not exceed 1.0 l/s.

Ground waters of friable fragmental sediments are associated mainly with the alluvium of river valleys, where they move in the streams form up to 1–2 km wide and up to 40–50 m thick. Groundwater flows at a depth of 0.5–10 m. Groundwater are fed by the filtration of surface waters, atmospheric precipitation and inflow of fractured waters.

Groundwater of modern alluvial sediments is stretched by narrow strip along channels of the rivers Shar and Kokpekty at the depth of 0.5–2 m, more rarely at the depth of 6–15 m. Pressurized waters come out at the lower reaches of the river Kokpekty. Mineralization of groundwater is different (from 250–500 to 1300 mg/l), composition is predominantly hydrocarbonate-sulfate. Average module of the Kalba underground flow is 1.5 l/s from 1 km^2.

Thus, the East Kazakhstan has fairly dense network of rivers and lakes, it is rich in underground waters, and mountain systems of the Southern Altai are centers of modern glaciation. Waters are closely interrelated with the climate, soil and vegetation cover and other components of nature and have direct impact on them. Different conditions for the formation of entire geographic flow, which is based on surface and underground drain develops under the influence of such technogenesis factors as the uneven location of river network in the flat part of East Kazakhstan, dense river network and high watercourse of the Altai mountain system, which contributes to the pollutants removal of natural waters and soils beyond the boundaries of industrial regions. All this have caused the differences in region natural potential and its ecological situation, requiring detailed geoecological studies of the East Kazakhstan geosystems with increased man-made loads.

1.7 Soil and Vegetation Covers

The East Kazakhstan is partially located in steppe zone, in subzone of dry fescue-feather grass steppes and semi-desert and desert zones. Features of latitudinal zonation are quite clearly manifested in foothills, as well as in deep and wide basins. Thus, 70% of the Zaisan basin area occupies desert and semidesert type of vegetation. Vegetation cover here is very rarefied. Here, the most diverse types of vegetation and soils are observed (Nekhoroshev 1932; Kuminova 1960; Revushin 1988).

The most diverse types of vegetation are observed here, from desert solo communities of the Kazakh Upland areas to taiga forests of north-east Altai and tundra associations of its highlands.

Brown desert-steppe soils with low humus content, only 1–1.2%, were formed in the belt of the South Altai low mountains and Zaisan depression in the north, in conditions of arid climate, and surface, as a rule, is strongly damaged by constant dry winds blowing round year (Flora Vostochnogo Kazahstana 1991; Popov 1940; Troitsky 1930).

This types of soils are replaced by chestnut and light-chestnut soils (Kurshym, north of Zaisan and Tarbagatai districts) typical for arid areas of lowlands, foothills, intermontane depressions and upland relief of the South Altai mountains. Thickness of humus horizon is 25–30 cm. Distribution zones of chestnut and light chestnut soils are called zones of risky farming, without an irrigation system, agricultural crops are not grown on them (Kolkhodzhayev 1974; Sokolov 1977, 1978).

At an altitude of 600–800 m, light chestnut soils are replaced by dark chestnut soils (Ulan and Kokpekty districts), which occupy the foot of the South Altai and Sauyr mountains. They have a well-defined humus horizon of brown color up to 40 cm thick. They are used as arable land for farming, where solid wheat varieties are grown.

Black soils are common in the foothills, low mountains and intermountain valleys of Rudny, South Altai and Kalba up to heights of 1500 m. Black soil plains are the best arable lands in the region and have a powerful humus horizon from 40 to 90 and 120 cm. They are the main agricultural zone, where many valuable agricultural crops. Mountain black soils are used as valuable pastures and hayfields (Makhambetova 1998; Matusevich 1934; Mukhlya 1939; Glazovskaya 1945).

Black soils are replaced by mountain gray forest and mountain meadow soils, which occupy only the north and south slopes of the Kalba middle mountains, Rudny, South Altai and Sauyr. They are characterized by layer of forest litter and thickness of the humus horizon up to 30–40 cm, they have enough moisture. These soils are occupied by aspen-birch, pine forests or fir-cedar taiga and represent valuable forestry lands.

High parts of the smoothed Alpine relief of the Rudny and South Altai, which are above 2000 m, are occupied by mountain-tundra soils. They have very low power, up to 30–40 cm, and they are rich in peat. They are characterized by natural vegetation, squat alpine grass and creeping shrubs. And there are stony placers and glaciers above mountain tundra soils (Mukhlya 1939; Glazovskaya 1945).

There are mainly steppe fescue and feather grass, fescue vegetation in the Priertis lowland. Ribbon forests interspersed with stripes of grass and shrub vegetation stretched from the north-east to south-west shows the peculiarity of steppes of the river Ertis right bank.

Steppe feather grass, fescue-grass vegetation are common on slopes of the Kalbin Range up to heights from 600 to 900 m. Belt of steppes is replaced by forest belt at altitudes from 900 to 1000 m in the central part of ridge. Subalpine meadows can be spotted at altitudes more than 1100–1200 m above the forest (Pochvyi Kazahskoy SSR 1968).

Forests are most widespread in the west and especially north regions of the Altai. Dominant part of the East Kazakhstan territory is occupied by aspen-fir black taiga, alternating at fir-trees between 1200 and 1700 m with fir-and-cedar moss forests. All foothills are occupied by forest-steppe vegetation in the North-West Altai. Coniferous-larch forests, represented by birch, larch, pine, cedar are located there at altitudes from 400 to 500 m.

There are wormwood gray, wormwood seliterane, prostrate summer cypress, fescue furrowed, and giant fleece, kermek gmelil in the sandy and clayey deserts of the Zaisan basin. Among shrubs there are white-barked calligonum, salt tree, multiflorate tamarisk, sand-grass saxaul. Sedge-meadow vegetation and reed thickets are formed on the river valleys.

Steppe is located in the low-mountain belt. Steppes pass into the taiga belt in middle reaches. There is an alpine belt of subalpine, alpine meadows and mountain tundra above them. Even higher, there is belt of eternal snows, glaciers and rocks along vertical zoning.

According to the natural conditions, ridges of Sauyr and Tarbagatai occupy an intermediate position between the mountains of Southern Siberia (Altai) and mountains of Central Asia (Tyanirtau). There are Siberian larch and Tien-Shan spruce in Sauyr, and its low mountains are occupied by semi-deserts (Rastitelnost Kazahstana 1941).

Dry semi-desert fescue-feather grass steppes extend on the north slope of the Tarbagatai ridge. They go into alpine meadows. Steppes are dominated by feather grass, fescue, cheegrass, short stalk and grass like fragrant onion, serpentine, stepless stone, etc. This type of vegetation is characteristic for desert steppes on the Mongolia territory. Shrub steppes of dog rose, twigs, buckthorns, currants are common on the south slope of the Tarbagatai ridge. There are small areas of forests from apple tree of Sivers, poplar and aspen on the gorges. There are tall grass meadows from catchment area, carthamoid rhapontic and other Siberian species of herbaceous plants.

Steppe zone takes large area in the territory. The steppe zone passes in form of fairly wide field along the north, west and south foothills of Rudny, South Altai and Kalba, along vast river valleys in the form of grassy-motley grass and feather grass black soil steppes. Foothill steppes are characterized by an abundance of herbage, height and dense grass cover, since they occupy the most moisturized zones part.

Meadow steppes are replaced by shrubby and feather-grass with thickets of honeysuckle, dog rose, barberry, meadowsweet, yellow pea shrub in areas with moderate moisture. Close gorges of the mountains northern slopes are covered with bird cherry, dog rose, meadowsweet, hawthorn.

Steppe vegetation of the Middle Asian form penetrates the Zaisan depression and the Ertis valley into depths of the South Altai, along the river valleys and dry corridors. These steppes are also quite diverse. They are dominated by feather grasshopper, fescue, and from bushes like spiraea and acacia.

Forest steppe belt is transitional. It is especially typical for the South Altai. Mixed and larch forests grow there, and there is steppe vegetation on the south slopes and intermountain hollows. The forest-steppe belt looms less clearly in the Rudny Altai.

Forest belt occupies about 40% of region territory. Upper limit of forest massifs is formed under the influence of heat and moisture factors. Five main coniferous species, which are spruce, fir, pine, larch and cedar, form a forest belt, depending on the terrain, climatic and soil conditions.

While moving to the south and south-east, total amount of precipitation decreases, and air dryness increases. As a result, dark coniferous breeds are replaced by light coniferous species. Various shrubs, meadow-walled and alpine grasses embed in the forest.

There are original fir forests, black taiga, in the Rudny Altai which is rich with its rainfalls. The main coniferous species in the South Altai are Siberian larch, spruce, fir, cedar with an admixture of birch and aspen. They form a continuous belt, which is bordered by rocky outcrops and riverbeds.

There is a big variety of forest landscape combinations in the forest belt, starting with mixed spruce-fir, spruce-larch to spruce-birch, aspen-birch and coniferous-larch forests.

Black forests (taiga) are severe and humid north-east part of the Rudny Altai, which is typical mountain taiga region. Black forests are combination of fir, Siberian spruce, cedar, and larch. They are joined by aspen and birch. Undergrowth usually consists of bird cherry, viburnum, mountain ash, with a high herbaceous cover. Aspen-birch and bird cherry groves occupy wide forest valleys, and form lush groves. They gradually merge with the dark coniferous taiga on mountains.

Above 1600 m, fir and spruce retreat, and cedar becomes the dominant species. It forms pure plantations—cedar forests.

In the South Altai, cedar forests also form a continuous belt. In the drier parts (Naryn Ridge), they are similar to Park Forest with an admixture of birch and aspen. Trees seem to be drowned in luxurious subalpine meadows.

The main forest-forming species in the South Altai is Siberian larch. It rises to height of 2000 m and prefers not very steep, moderately wet slopes of open valleys.

Pine forests are found on the Rudny and South Altai, but they are most typical for the Kalbin ridge. There pine forests occupy large granite massifs. From grasses prevail bluegrass forest, fescue furrowed, hawk umbrella, resin drooping. There are also endemic species listed in the Red Book—venus shoe, peony steppe, almond Ledeburovsk, etc.

Nature monuments—relics of the Tertiary period—Kaindy forest and the Sinegorov fir grove—islets of the past epochs have been preserved on the Kalbin ridge.

High mountain belt, in the Rudny and South Altai it is characterized by total lack of forest. In these zone mountain meadows, thickets of round-billed birch, moss-lichen and stony tundra are widespread. Snow cover is missing only for two or three summer months there. Highland belt is divided into two lines—mountain-meadow and mountain-tundra. Strips are not sharply delimited. One wedge into the other, transitional groups is formed at the points of contact.

Mountain-meadow belt of the Rudny and South Altai is the subalpine and alpine meadows. Subalpine meadows dress slopes of mountains in woodlands strip and slightly above it. They have an external similarity with forest meadows. Due to the high air humidity, plants reach a considerable height as 1.5 m.

The moss-lichen mountain tundra is located on more humid slopes, along the bottom of wide valleys and in upper reaches of the Rudny and South Altai rivers. This type is characterized by continuous moss and lichen cover. There are undersized, dense shrub thickets of round birches and willows. There are tundra "sogra"—marshy, swampy, grassy and mossy spaces.

The crushed-lichen mountain tundra occupies the higher slopes of Rudny, South Altai and Tarbagatai. The stony mountain tundra occupies the highest parts of high altitude belt of the Altai and adjoins snow line. Here settle more hardy squat plants —bluegrass, tonkonog and other cereals. The rest of space is bare rocks, stony rises, snowfields. The highest parts of the Rudny, South Altai and Sauyr ridges occupy the belt of eternal snows, glaciers and rocks. Small areas of glaciers and snowfields are confined to the highest point of Altai—Mount Belukha (4506 m) (Fizicheskaya geografiya Kazahskoy SSR 1959; Kazahstan. Obschaya fiziko-geograficheskaya harakteristika Kazakhstan. Obschaya fiziko-geograficheskaya harakteristika 1969; Chupakhin 1970, 1987; Geldyeva 1992).

Flora of the East Kazakhstan region includes more than 3500 species of higher plants (from 5700 species of higher plants in Kazakhstan). The richest species of family are Compositae and cereals. There are relatively many species in such families as butterflies, rosaceous, cruciferous, buttercup, clove, sedge.

Diversity of natural conditions of the East Kazakhstan was reflected in richness of its fauna. There are 109 species of mammals (63% of the species composition of the republic fauna), 373 species of birds (83%), 22 species of reptiles (45%), 5 species of amphibians (49%), 37 fish species (32%), more 15,000 species of insects. Each landscape zone, high altitude belt has its own original set of animal species that live there.

The second edition of the Red Book of Kazakhstan, published in 1991, introduced many species of animals that live in the East Kazakhstan and endangered or have greatly reduced the number and area of their habitats.

Thus, 38 species of birds, 13 species and subspecies of mammals, and 4 species of reptiles are included on its pages. And such species as kulan, saiga, antelope, gazelle, and local beaver subspecies disappeared from the territory irretrievably.

Features of latitudinal zone are quite clearly manifested in the foothills, as well as in deep and wide basins. Thus, 70% of the Zaisan basin area occupies desert and semi-desert type of vegetation. Here, vegetation cover is strongly rarefied (Revushin 1988; Flora Vostochnogo Kazahstana 1991; Popov 1940).

Thus, the soil and vegetation cover of the East Kazakhstan has zonal character of distribution; in the mountain basin part, there is a vertical change of soils, in close dependence on the height of mountain ranges, forms of relief and orientation of the ridge slopes. Distinctive ecological properties of soils in the East Kazakhstan are the complexity (alternation of different types on soils and soil varieties), the consequence of the terrain heterogeneity, the climate aridity, the geological structure and soil-forming rocks, which determines the agricultural land use. Unsustainable economic activities of land use have resulted in such consequences as wind and water erosion, low natural fertility requiring detailed, targeted development of nature protection measures aimed at the protection and biota restoration.

1.8   Landscape Structure

Ecological problems of the East Kazakhstan are largely determined by features of natural landscapes, within which economic activities are carried out (Chigarkin 2003).

Systematics of landscapes presupposes the grouping of landscapes into species, gender, types of landscapes according to their structural, genetic, and functional generality. In the process of typological classification, the most important properties of landscapes are established.

Taxonomic rank of geosystems classified and mapped on a small scale corresponds to the geographical landscape, i.e. units occupying hierarchical position at the junction of regional and local geosystems. This means that landscapes are collection of regularly repeated in the space of local geosystems-tracts, subtracts, faces, united by common par agenesis and lateral real-energy relationships. Genetic unity of geographical landscapes is emphasized by the fact that they are always geographically associated with genetic types of terrain or relief—derivatives of morpholithogenesis (Geldyeva 1992, 1971; Kolomyts 1987).

For the systematics of the Ertis river basin landscapes, structural and genetic classification proposed by A.G. Isachenko and V.A. Nikolayev was used (Armand 1979; Nikolaev 1979; Sochava 1963; Isachenko 1972).

On landscape map of the Ertis river basin, we have defined the classes of flat and mountain landscapes. In the plain landscapes, three subtypes are distinguished (dry-steppe, semi-desert and desert). Each subtype is divided into genera: accumulative, denudation (from 1 to 20). In the class of mountain landscapes, three subclasses and four genera are distinguished (from 20 to 70). In addition, we have identified intrazonal valleys and floodplain landscapes (from 71 to 75). In total, 3 types of landscapes were identified on the landscape map of region, 5 genera and 19 landscape species (from 1 to 19 landscapes). In the class of mountain landscapes— 6 types, 3 subclasses, 12 genera and 51 kinds of landscapes (from 20 to 70 landscapes). Intrazonal ones are represented by 5 species of valley hydro orphic landscapes (Fig. 1.1), (Appendix A).

Thus, by analyzing the natural factors of functioning, it can be concluded that landscapes of the East Kazakhstan, experiencing anthropogenic load, represent functional-integral geographic system with single tectonic-gravitational structure that has unified factors of landscape formation and development that are unstable in time and space. Since all physical-geographical processes are subject to unidirectional horizontal flows of liquid drain and lateral gravitational flows of solid flow directed from watershed to erosion base.

Consideration of ecological properties of landscapes should be based on the principle of nature protection, which assumes comprehensive evaluation of all interconnected natural components and technogenic factors of their constituents.

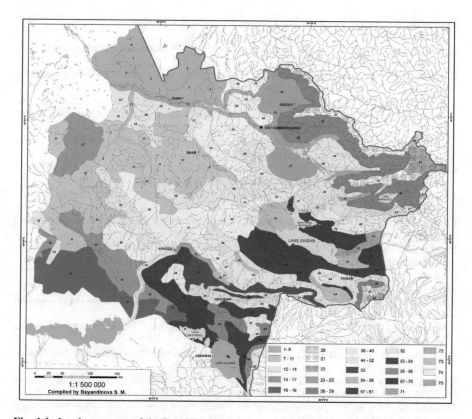

Fig. 1.1 Landscape map of the East Kazakhstan

References

Akhmedsafin UM (1971) Resursyi podzemnyih vod i gidrologicheskie prognozyi v zone perebroski chasti stoka sibirskih rek v Kazahstan (Resources of underground waters and hydrological forecasts in transfer zone on drain part of the Siberian rivers to Kazakhstan). Nauka, Alma-ata, p 441

Armand AD (1979) Nauka o landshafte (Science about landscape). Moscow, pp 16–30

Atlas of the Kazakh SSR (1982) Publ. SDGS at CM of the USSR, Prirodnyie usloviya i resursyi (Nature conditions and resources). Moscow, p 82

Beysenova SA (1982) Fiziko-geograficheskie issledovaniya prirodyi Kazahstana 1917-1941 (Physiographic researches of the Kazakhstan nature 1917–1941). Kazakhstan, Alma-ata, p 176

Bublichenko NA, Vorobyov YY, Ivankin PF (1972) Printsipyi i metodyi prognozlrovanlya mednokolchedannogo i polimetallicheskogo orudneniya (na primere Rudnogo Altaya) ((Forecasting principles and methods of chalcopyrite and polymetallic ore (on example of the Rudny Altai))). Nauka, Moscow, p 256

Bublichenko NA (1936) Osnovnyie tektonicheskie linii Rudnogo Altaya, Bolshoy Altay (Main tectonic lines of the Rudny Altai and Big Altai). Academy of Sciences of the USSR publ, Tectonics, p 229

Chikhachev PA (1974) Puteshestviya v Vostochnyiy Altay (Travel to the East Altai). Nauka, Moscow, p 360

Chigarkin AV (1974) The main problems of landscape science and nature protection in Kazakhstan. Kazakh National University, Alma-Ata, p 135

Chigarkin AV (1995) Geekologiya Kazahstana (Geoecology of Kazakhstan). Al-Farabi Kazakh National University, Alma-ata, pp 22–37

Chigarkin AV (1996) Geoekologicheskoe rayonirovanie i ekologicheskaya narushennost geosistem Kazahstana (Geoecological division and ecological fracturing of Kazakhstan geosystems). Reporter of Al-farabi Kazakh National University, Geography series № 6, publ. 36, pp 19–27

Chigarkin AV (2003) Geoekologiya i ohrana prirodyi Kazahstana (Geoecology and nature conservation of Kazakhstan). Kazakh universities, Almaty, pp 41–55

Chernykh S (1971) K istokam chernogo Irtyisha (ob issledovaniyah V.V.Sapozhnikova) (Sources of Kara Ertis (about V.V. Sapozhnikov's researches)). Rudny Altai, 4 Sept 1971

Chupakhin VM (1970) Prirodnoe rayonirovanie Kazahstana (Natural zoning of Kazakhstan). Nauka, Alma-ata, pp 38–49

Chupakhin VM (1987) Vyisotnozonalnyie geosistemyi Sredney Azii i Kazahstana (High zonal geosystem of Central Asia and Kazakhstan). Nauka, Alma-ata, pp 17–23, 57–63

Egorina AV (2004) Barernyiy faktor v razvitii prirodnoy sredyi gor (Barrier factor in development of the mountains environment). Auto report of geography sciences dissertation work, Barnaul, p 47

Flora Vostochnogo Kazahstana (Flora of the East Kazakhstan) (1991) Nauka, Alma-ata, p 441

Fizicheskaya geografiya Kazahskoy SSR (Physical geography of the Kazakh SSR) (1959). Kazakh state study pedagogical publication, Alma-ata, pp 44–52

Geldyeva GV (1971) Veselov LK. Landshaftnaya karta Kazahskoy SSR (Landscape map of the Kazakh SSR). SA Reporter KAZSSR № 11, pp 13–19

Geldyeva GV (1992) Veselov LK, Landshaftyi Kazahstana (Landscapes of Kazakhstan). Nauka, Alma-ata, p 173

Geologicheskaya izuchennost SSSR. Kazahskaya SSR (Geological study of the USSR. Kazakh SSR) (1974) the East Kazakhstan Kazakhstan, Alma-ata, p 118

Glazovskaya MA (1945) Poyasnitelnaya zapiska k pochvennoy karte (Explanatory note to the soil map). M: 1:1000 000 the EKR In: Gerasimov IP (ed), Alma-ata, p 12

Isachenko AG (1972) Puti sinteticheskogo izobrazheniya prirodnyih kompleksov, izmenennyih deyatelnostyu cheloveka (Ways of synthetic image on natural complexes changed by human activity) Synthesis problems in cartography. MSU, Moscow, pp 13–34

Kaletskaya MS, G.A. Avsyuk, S.N. Matveev (1945) Goryi Yugo-Vostochnogo Kazahstana (Mountains of the Southeast Kazakhstan). Alma-ata, p 241

Kassin NG (ed) (1941) Geologiya SSSR (Geology of the USSR). State publication of geological literature, the East Kazakhstan, p 312

Kassin NG (1947) Materialyi po paleogeografii Kazahstana (Materials on paleogeography of Kazakhstan). SA KAZSSR, Alma-ata: 1947, p 318

Kazakhstan. Obschaya fiziko-geograficheskaya harakteristika (Kazakhstan. General physiographic characteristic) (1969) Academy of Sciences of the USSR, Moscow, pp 17–21, 31–42

Katalog lednikov SSSR (Glaciers catalogue of the USSR) (1969) T.15, vol 1, Hydro meteorological publication, Leningrad, p 85

Klink V (1976a) Ot Zmeinogorska do Riddera (k 120-letiyu issledovaniya Rudnogo Altaya P. P. Semenovyim Tyan-Shanskim) ((From Zmeinogorsk to Ridder (to the 120 anniversary of Rudny Altai research by P.P. Semenov Tian-Shan))), Leninogorsk truth. 21 Aug 1976

Klink V (1976b) Glazami P.P. Semenova Tyan-Shanskogo (iz istorii issledovaniya Rudnogo Altaya) ((As viewed by Semenov Tian-Shan (from history of Rudny Altai research)). Rudny Altai, 1 Oct 1976

Kurochina LY, Makulbekova GB (1996) Ecosystem changes in the northern (Kazakhstan) area of the former Aral Sea (Priaralie). In: Micklin PP, Williams WD (eds) The Aral Sea Basin. NATO ASI Series (Series 2. Environment), vol 12. Springer, Berlin, Heidelberg

Kolkhodzhayev MK (1974) Pochvyi Kalbyi i prilegayuschih territorii (Soils of Kalby and adjoining territories). Alma-ata, p 664

Kolomyts EG (1987) Landshaftnyie issledovaniya v perehodnyih zonah (Metodologicheskiy aspekt) (Landscape researches in transitional zones (Methodological aspect)), Nauka, Moscow, p 117

Kuminova AV (1960) Rastitelnyiy pokrov Altaya (Vegetation cover of the Altai). Academy of Sciences of the USSR, Novosibirsk, p 431

Kyzykbayev T (1964) Slavnyie stranitsyi (o prebyivanii G.N. Potanina v Ust-Kamenogorske) (Nice pages (about G.N. Potanin's stay in Ust-Kamenogorsk)). Reds banner, 10 Oct 1964

Makhambetova ZU (1998) Pochvyi Kazahstana (Soils of Kazakhstan) 1991-1997: bibliographic index. MES RK, Almaty, p 217

Matusevich SP (1934) Pochvennyiy pokrov Kazahstana (Soil cover of the Kazakhstan. Regional publication). Alma-ata-Moscow, p 771

Mukhlya AV (1939) Serozemnyie pochvyi i pochvyi gornyih oblastey Kazahstana (Sierozemic and mountain area soils of the Kazakhstan). Collection on the Kazakhstan soil cover, Alma-ata, p 28

Nekhoroshev VP (1932) Drevnee oledenenie Altaya. Trudyi po izucheniyu chetvertichnogo perioda (Ancient freezing of the Altai. Works on studying of the quaternary period). Academy of Sciences, Moscow, p 471

Nekhoroshev VP (1934) Kratkiy geologicheskiy ocherk territorii Bolshogo Altaya (Short geological sketch of the Big Altai territory). Kazakhstan, Alma-ata, p 420

Nekhoroshev VP (1936) Novyie dannyie po geologii Bolshogo Altaya (New data on the Big Altai geology) Kazakhstan, Alma-ata, p 510

Nikolaev VA (1979) Problemyi regionalnogo landshaftovedeniya (Problems of regional landscape studying). Academy of Sciences of the USSR, Moscow, pp 26–38

Obruchev VA (1927) K voprosu o tektonike Altaya (To question of the Altai tectonics). Geol. Reporter № 4–5, pp 17–24

Obruchev VA (1963) Moi puteshestviya po Sibiri (My travel across Siberia). Moscow, p 287

Obschaya harakteristika geomorfologii (Total characteristic of geomorphology) (1947) Geology of the USSR, the East Kazakhstan, Geological description, Moscow, p 214–218

Przhevalsky NM (1948) Iz Zaysana cherez Hami v Tibet i verhovya Zheltoy reki (From Zaisan through Khami to Tibet and upper courses of the river Yellow). ASBJP, Moscow, p 406

Popov MG (1940) Pochvenno-rastitelnyiy pokrov Kazahstana (Soil and vegetation cover of the Kazakhstan). Academy of Sciences of the USSR, Moscow, p 572

Pochvyi Kazahskoy SSR (Soils of the Kazakh SSR) (1968) Vol 10: Semipalatinsk region, Alma-ata, 1968, p 631

Rastitelnost Kazahstana (Vegetation of Kazakhstan) (1941) SA KAZSSR, Moscow, p 618

Revushin AS (1988) Vyisokogornaya flora Altaya (Mountain flora of the Altai). Tomsk University, Tomsk, p 284

Sarsenbayev MH (2001) Gidrologo-ekologicheskie problemyi orosheniya v Yuzhnom Pribalhashe (na primere risovyih zemel) ((Hydrologic-environmental irrigation problems in the South Pribalkhash (on example of rice lands))). Kazakh Universities, Almaty, p 196

Sarsenbayev SM (ed) (1971) Vodnoe hozyaystvo Kazahstana (Water management of Kazakhstan) Kaynar, Alma-ata, p 318

Sedelnikov AN (1898) Geobotanicheskoe opisanie Naryimskoy dolinyi na Altae (Geobotanical description of the Narym valley in Altai). Nauka, Moscow, p 133

Scientific and application-oriented reference manual on climate of the USSR (1989) publ. 18: Kazakh SSR. Book 2, Hydro Meteorological Publication, Leningrad, p 440

Shchurovsky GE (1980) Geologicheskie opisaniya po Altayu (Geological descriptions across the Altai). Nauka, Moscow, p 325

Shcherba GN, Dyachkov BA, Stuchevsky NI (1998) Bolshoy Altay (geologiya i metallogeniya) (The Big Altai (geology and the metalgenius)). Book 1, Geological structure, RHS SCADT RK, Almaty, p 304

Shcherba GN, Dyachkov BA, Stuchevsky NI (2000) Bolshoy Altay (geologiya i metallogeniya) (The Big Altai (geology and the metalgenius)) Book 2: Metalgenius, RHS SCADT RK, Almaty, p 304

Sochava VB (1963) Opredelenie nekotoryih ponyatiy i terminov Fizicheskoy geografii (Definition of some concepts and terms on Physical geography). In coll. report of Siberia and the Far East geographical university № 3, vol 3,1963, p 50–59

Sokolov AA (1977) Obschie osobennosti pochvoobrazovaniya i pochv Vostochnogo Kazahstana (General features of soil formation and the East Kazakhstan soils). Alma-ata, p 334

Sokolov AA (1978) Pochvyi srednih i nizkih gor Vostochnogo Kazahstana (Soils of the East Kazakhstan average and low mountains). Nauka, Alma-ata, p 224

Spassky GI (1809a) Puteshestviya po Tigiretskim belkam (Travel on Tigiretsk snow covered mountain peak). Nauka, Moscow, p 234

Spassky GI (1809b) Puteshestviya po Yuzhnyim Altayskim goram (Travel across the South Altai mountains). Nauka, Moscow, 1809, p 317

Tsybulsky VV (1988a) Nauchnyie ekspeditsii po Kazahstanu (Scientific expeditions across Kazakhstan). Kazakhstan, Alma-ata, p 112

Tsybulsky VV (1988b) Po gornyim massivam Altaya i dolinam velikih russkih rek (All along mountain massif of the Altai and great Russian river valleys). Nauka, Moscow, pp 74–97

Troitsky V (1930) Materialyi k harakteristike pochv Vostochnogo Kazahstana (Characteristic materials of the East Kazakhstan soils.). Semipalatinsk, p 367

Uteshev AS (ed) (1959) Klimat Kazahstana (Climate of Kazakhstan). Hydro Meteorological Publication, Leningrad, p 338

Tronov MV (1949) Essays on the glaciation of Altai. Geografgiz, Moscow, p 290

Veyts BI (1959) Mineralyi Rudnogo Altaya (Minerals of the Rudny Altai). Kazakhstan, Alma-ata, p 488

Vilesov EN (1977) Sovremennoe oledenenie Zailiyskogo Alatau i ego svyaz s orografiey (Modern freezing of Zailiysky Alatau and its communication with orography). Coll.: Dynamics of highlands natural processes. Academy of Sciences of the USSR, Leningrad, pp 60–76

Vodnyie resursyi Kazahstana (Water resources of Kazakhstan) (1957) SA KAZSSR, Alma-ata, p 217

Chapter 2
Technogenic Conditionality in Development of Geosystems in East Kazakhstan

2.1 Allocation Methods of Technogenic Geosystems

2.1.1 Theoretical Substantiation for the Organization of Geosystems in East Kazakhstan

Geoecology or landscape ecology is a science located at the intersection of two sciences as geography and ecology, considering the ecology of geographic systems or environment of human activity within the boundaries of natural and territorial complexes of different rank.

Theoretical foundations of genecology are determined by the regularities in the development of geographical envelope and the Earth biosphere. The biosphere, being a part of the geographical envelope, differs from it by less power and high concentration of life (Chigarkin 2003; Dzhanaleeva and Bayandinova 2004).

Depending on the methodological approaches and scientific and practical purposes, the objects of environmental studies can be either geosystems (in biology) or geosystems or landscapes (in geography). Geosystem or landscape was chosen, since spatial differentiation of the geographical envelope is due to combination of many natural components, the main object of which is the geoosystem, and spatial division of the biosphere is based on its differentiation by our object of study (Nekhoroshev 1934; Grebenshchikov and Tishkov 1986).

Structurally, landscapes are analyzed in two ways: on the one hand, as natural complexes consisting of local morphological units, on the other, as elements of larger regional units (provinces, natural zones, countries). In other words, the internal and external structure of landscapes is taken into account.

In view of the fact that geographical landscape is internally non-uniform, the question arises which of the morphological parts, their components, first of all should be taken into account, assuming them as the basis of classification. Here the notions of dominant, subdominant and other subordinate morphological units of landscape. Properties of the dominant tracts in the landscape are recognized as the

© Springer Nature Singapore Pte Ltd. 2018
S. Bayandinova et al., *Man-Made Ecology of East Kazakhstan*,
Environmental Science and Engineering, https://doi.org/10.1007/978-981-10-6346-6_2

main subject of co-positional typological analysis. Subdominant tracts can also provide valuable information for landscape diagnosis, but they are taken into account, secondarily.

Structural and genetic classification of landscapes can not fail to take into account their regional division. The influence of specific geographical position (place) on the earth's surface always affects the history, genesis and modern structure of any landscape. For this reason, the elements of regionalism are inevitable in landscape-geographical classifications. They are given a regional-typological character (Demek 1977; Krauklis 1979, 1989).

The classification of multistage landscapes consists of hierarchy on taxa, from the top to bottom more and more concretizing the typological characteristics of geosystems. Each taxonomic level corresponds to the strictly defined classification feature, in formal logic designated as the unity of basis for concepts' division.

The set of classes and subclasses of landscapes that follow the classification ladder below depends on the relief morphstructure. Class of flat landscapes in region completely represents the outlying terraces, and the mountainous ones—on low-mountain, middle-mountain and high-mountain landscapes. The division of landscapes into classes and subclasses reflects one of the most important aspects of landscape envelope—its layering, caused by the geotectonic movements of the earth's crust. It is directly reflected in the longitude of landscape, which, in turn, affects water regime and geochemical specificity of geosystems.

In the class of flat landscapes, accumulative auto orphic and semi-hydro orphic geosystems develop. For the elevated plains, auto orphic denudation landscapes are characteristic, for the low drainage ones—neo-elluvial (paleohydromorphic), for the undrained plain interfluves and bottoms of river valleys and lake basins—semi-hydromorphic and hydro orphic. In small hill landscapes the trans-elluvial geosystems are dominating.

Manifestation of zonal or intrazonal features of landscape directly depends on the degree of automorphicity-hydromorphism. Steppe placers are always auto orphic. They are standard of natural zoning. At the same time, hydro orphic lowland positions are usually occupied by intrazonal—marsh, meadow, solonetzis-saline geosystems. However, we emphasize that intrazonal landscapes are also zonal, but their zonal nature is distorted by increased hydromorphism (ground, muddy, and floodplain). As a result, in a single family and in one class of landscapes, both zonal and intrazonal geosystems can simultaneously develop in the conditions of different water regimes, forming mosaic of landscape types.

The type of landscape is one of the main classification units. The basis of the concepts selection is soil-geobotanical characteristics at the level of soil types and classes of plant formations. If we confine ourselves to the early stages of auto orphic landscapes aggregate, then the landscape type is fairly accurately geographically correlated with the natural zone within single physical-geographical country. However, next to them, in the same zone, but in hydro orphic positions, one can identify saline-meadow, marsh-meadow and other intrazonal types of landscapes of the same family and class. Together with auto orphic-zonal, they form the landscape structure of the Altai mountain ranges (Sochava 1978; Solntsev 1981; Isachenko 1972, 1981; Dzhanaleeva 1986, 1993).

The types of landscapes consist of subtypes. The most obvious classification characteristics of subtypes for auto orphic landscapes are their inherent soil subtypes and subclasses of plant formations. Subtypes of zonal landscapes correspond territorially to adjacent subzones.

Below in the hierarchy of typological taxa follow genera and subgenera of landscapes, characterized by the geological and geomorphologic features. Morphology and genesis of relief is an indicator of landscape kind; litho logical characteristics of surface sediments are indicator of the landscapes subgenus. It is expedient to divide the genera of landscapes into two large aggregates: (1) landscapes of interfluves; (2) landscapes of river valleys and lake basins.

Particular attention is drawn to the landscapes systematic position of high—third (Middle Pleistocene) above-floodplain terraces of river valleys in flat areas. According to hypsometric position, and the most important auto orphic, neo-alluvial nature, they practically do not differ from the ancient alluvial and lacustrine-alluvial plains of the interregional, and therefore are included in the latter composition. In composition of valley and lake-hollow landscapes of the plains, there are natural geosystems of low—the first and second (late Pleistocene) above-floodplains and flood plains, always hydro orphic to some extent. Another thing is the mountain-congested areas of the Ertis river basin, where the entire complex of over pristine terraces, including high terraces, is strictly localized in the river valleys (ancient and modern), and according to the classification it belongs to the category of valley natural geosystems.

At the level of landscapes subgenus, the litho-factor plays an important role. It defines the diversity of the steppes edaphic variants—from loamy and loess (pelitophytic) to psammophyte, petrophytic, halophytic, calciferous, etc. According to G.N. Vysotsky, it is customary to consider the forest-loamy variant of steppe as the steppe model zone. Sandy or stony water dividing plain can not be recognized as steppe placer, since they represent the corresponding lytho edaphic steppe variant. The impact of lytho-edaphic factor sometimes becomes so strong, that it leads to emergence of extra zonal landscapes among steppes.

One of the last stages of hierarchical ladder on typological taxa is the type of landscapes. It represents a set of natural complexes, similar in composition to the dominant tracts in them. Of course, such structural proximity assumes the commonality of their evolution and genesis.

Let's remind that under laws of hierarchy, the characteristics of all higher taxons have the defining value for subordinate. As a result, each type (morphological option) of landscapes receives its own individual characteristic.

Retrospective analysis of theoretical concepts concerning landscape science of the last years claims the methodological geosystems approach to take place in studying the large physiographic regions. Systematic character of the most geographical envelope follows from the fact that communications between the components are stronger, than communications with the external environment.

Geosystem is an open system, exchanging substance and energy with environment. The super openness of geosystems river basin is caused by constant dependence of superficial drain elements on rainfall, which are various in different parts of pool. It defines also liquid intensity and firm drain of substances, having the uniform direction, and connected with difficult transfer processes of substance (an erosion, accumulation, etc.).

Transition from one spatially—large-scale level to another in the conditions of pool is followed by the high-quality reorganization of natural bodies, physical interaction ways and creates thorough geosystem. So, river basin geosystems from upper parts to delta are united with general process of the movement and transfer of the weighed particles, as it is one of genesis and modern functioning of geosystem factors.

One of the best conceptual decisions in system approach was identified by V.B. Sochava (Nikolaev 1978; Sochava 1978)—"geosystem can be found as a system of interaction between geographical spheres with hierarchical structure and functional similarity and unity of spatial communications". On basis of this definition the geosystems of all river basins can be considered as a united complex, geographically constructed, located as equals, but rather independently functioning simple and difficult natural systems, and the complexity principle is "implemented through identification of all taxons".

Structurally-functional links of geosystems in geosystem classification in the river basin rely on the principle of thorough systemacity. As the united geosystem, a river basin is the super difficult, exoregulated, impulsively dynamic geosystem limited to two special types of surfaces: threshold—vertical (for example, glacial zone) and contact—horizontal (river floodplain). When studying geosystems of internal drain it is necessary to consider, in our opinion, non-traditional component blocks, since besides lithogenous basis, differentiating factor is the superficial drain, also components of macro—and micro substratum levels of geosystems. We refer parameters of water and thermal balance, efficiency and productivity of phytoweight to them. We consider geosystems as united midland, formed by drain rivers, as par genetic and pardynamic complexes in conditions of amplifying moistening deficiency owing to natural and anthropogenous factors. These natural complexes develop under the influence of two interdependent, leading factors of differentiation—lithogenous basis and drain. These and other physiographic conditions, forming a river basin, allow defining the region as united megageosystem.

Super complexity, impulsive dynamism, exoregulating feature of low rank geosystems at all levels of geosystem organization from all macrogeosystem according to V.N. Solntsev, are proved by three main types of geosystem structures: (1) vector, (2) cellular, (3) is potential types of geosystems structures. Such polystructure of geosystems can be described as existence in complex organizations pools of circulating, radiation characteristics which are fundamental in system of interaction—the plain—the mountain, and owing to existence of "barrier effect", greenhouse effect, and also other difficult geographical processes arising in the conditions of one macrogeosystem with universally oriented geodrain.

Genesis of three structural types is connected with the types of physiographic processes happening in the pool: the first type of structure with external (sunlight) that is defined by inflow of solar energy, vertical currents, in general, caused by the area latitude, a northern and southern exposition, the kettle-hole and water separate provision of geosystems. The second type is connected with intrathecal (circulating). To this belongs geosystems of low-hill terrain and foothills subgeosystems. The third one is intra terrestrial (gravitational and tectonic processes), also geosystem of drain formation zones.

This concept generalizes a number of the long ago established geosystems representations: independence of zone and a zonal geosystems differentiation by Isachenko (1981), genetic relationship of high-rise and width zonality by A.D. Armande (Armand 1975), basin forms existence of geosystem orderliness by F.N. Milkov (Milkov 1967). Theoretical concept of links between these three types of geosystem structures is well lit in literature by Solntsev; etc. This position is also approved by A.Yu. Reteyum (Solntseva 1982) in the "monistic concept", where polystructural approach in researches of geosystems is interpreted. Macrogeosystem of the river basin is considered by us as the geosystems, which has territorial stability at the expense of a lithogenous framework with high plasticity of a biota, and connected with dynamics of superficial drain (Iberla 1980; Jeffers 1981).

In the conditions of East Kazakhstan river basins, usual lithogenous-petrographic characteristic of rocks is insufficient for their backbone role clarification. Structure and condition analysis of territorial rocks set bedding is necessary. A.N. Perelman (Reteyum 1975) put forward the idea about monolithic and heterogeneous geosystems, understanding under the second geosystems formed on various rocks: A.G. Isachenko (Perelman 1964; Isachenko 1974),—"geomorphologic complexes". According to N.A. Solntsev these representations are inevitably closed with the known definition of geosystems.

In contexts of river basins, there can be often met combination of Paleozoic metamorphic and intrusive breeds. Significant areas are occupied with friable quaternary deposits. Combination of various lithogenous-structural complexes defines mosaic alternation of radical and friable education sites that is one of mainstream fundamental factors.

Geological substratum, which is located below maternal breed, exerts impact on evolution of geosystems and has another time characteristic. In this sense the geological substratum is not seasonally functioning part of geosystem. Geological substratum in relation to concrete geosystem should be called the geological buildups, stable in characteristics within a year and, nevertheless, predetermining main features of a structure of the higher in the taxonomical row.

If geographical buildups are considered in the highest units as a direct functioning site, then for the lowest unit it acts as a background, condition, and substratum. The question of a lithogenous basis and substratum is connected with a vertical borders problem of physiographic units—particularly complex buildups.

Lithogenous basis combines local features of air, water, soil organisms, and mining masses, which do not have relations with modern geographical coordinates.

Geological structure has a significant effect on the form of valleys and longitudinal river profile, the structure of the alluvium, and on the stability of the river course. The modern tectonic movements influence placement of sites, which are limited and have free development of channel changes, prevalence of deep and side erosion.

Gravitational forces, being the indicator of substance and energy reorganization, determine the altitudinal zonality developed in subgeosystems of pool. Regularities of altitudinal zonality formation develop the geosystem tiers, representing extremely important regularity of the physiographic mountains differentiation. In the basins of the Altai rivers, mountainous geosystems are divided into three traditional tiers—low, average, high. These three main tiers reflect formation stages of a mountain construction, age of its separate parts, intensity of tectonic movements, as well as character of an exogenous partition. From here two conclusions follow: the first is phenomenon of a layering has to be the basis for geosystems differentiation; the second is layering changes of a mountainous terrain concerning the studied region involve changes of bioclimatic components.

In the geosystems of subdued mountains and low-hill terrain basin of the river Ertis, different genetic breeds are territorially interfaced. Besides, the contrast of the modern geomorphologic processes connected with a drain is especially big and obvious. The more physical distinctions between breeds making a complex, the sharper and more certain become the border of morphological structures and textures. Low-hill terrain and foothills of the Altai act in the form of an independent lithogenous basis and, being strongly average on material structure, make an impression about the created new lithogenesis—a substratum for the soil and biota. The role of superficial drain is obvious here: superficial washout and glacial waters created this relief as a result of bias. Foothills geosystem belongs to the class of mountain plains as in the tectonic plan these territories belong to Paleozoic age. Foothills are an initial step of a mountainous terrain morphological layering that finds reflection in allocation of the foothill geosystems subclass. Therefore, both a mountain part of the Ertis river basin, and the foothills of main Altai ridges are considered to be a united pargenetic geosystem.

In our researches exogenous processes play a large role in the intrasystem differentiation, since processes of a substance geographical drain (substance streams, underground drain, geochemical drain, and superficial drain), its character and force depend on work of a superficial drain.

Sediment processes of proluvial-alluvial material form a lithogenous basis of inflow courses concerning the above-mentioned rivers. This process happens every second and for many centuries, but as soon as at the same time at least one of lithogenous basis and geological substratum components will change, then the lithogenous basis acts as the differentiator (and sometimes the integrator) of geosystem structure. Thus, conservatism of lithogenous system is very relative in the conditions of the Ertis big and small river basin.

Within the framework of channel changes free development, weak stability of friable breeds composing a rivers bed and small alluvial course causes mainly a stream role in course processes (the stream operates the course). Such courses are the same for all zones of drain dispersion. In the conditions of breeds distribution, resisting to washout, which are rocky, connected, plastic, stream copes with the course. It is distinguishing for zones of drain transit. In the first case widely inundated courses are formed, in the second one there formed cut with valleys, following tectonic violations.

By virtue of channel, slope and ravine processes movement of deposits coming to the river can be carried out. Character of deposits and their power play an important role in functioning of geosystems concerning alluvial and deluvial-proluvial covers of all basin geosystems. The main part of deposits drain is formed on reservoir, representing a zone of drain formation. At the same time, share of the last, in general deposits drain of rivers reaches 60%, though from everything the material which is washed away from slopes on reservoir this size makes more than 20%. The most part of products concerning soil erosion settles at the bottom of slopes, which are defined as a drain feathering-out zone.

Channel processes, developing in the region have specific provincial features, and in general define the current state of geosystems. They are closely connected with backbone factors and by modifying, also change factors.

Channel processes are zoned, and in general also define zone signs of geosystems. Key parameters of geosystems are thermal and water balances, and efficiency correspond to laws of altitudinal zonality in the geosystems, functioning in formation zones and drain transit, also zone signs are clearly traced in feathering-out zones and drain dispersion. Such interconditionality is characteristic for all basins of the river Altai.

Channel processes happening in the Ertis river basin have more considerable interconditionality to high-rise belts, and less to line zone, which are primitive and muffled here.

Influence extent to the channel of physiographic factors depends on the channel deformations scale and on the sizes of channel forms. So, on the main drain of the Ertis river are featured by intrazonal geosystems, integrating in themselves influence of zone lines. In small rivers basin, channel mode is defined by lines and excludes an intrazoning. Often the intrazoning of inundated geosystems is subordinated to zone signs.

As indicators of major factors concerning channel processes, geological and geomorphologic conditions of channel formation, channel figuring processes and channel stability are accepted. Differentiation of the main types of channel process allows dividing geosystems into: (a) geosystems of the river and its basin flat part; (b) geosystems of Semi-mountain Rivers—in low-hill terrain and foothills; (c) geosystems of Mountain Rivers. In mountainous areas the altitudinal zonality of channel processes defining natural change of one type by another is evident.

In the geosystems, assigned to the main course of the Ertis river, genetically different water streams are revealed—not channel, flowing down from slopes; the streams, defining concentrated accumulation in the form of carrying out cones; constantly channel, forming river valleys during geological history. All of them play a different role in differentiation, functioning and dynamics of the geosystems concerning the studied region. The role of water streams in the geosystems development is stipulated not only by fact that the first two types of streams (not channel and temporarily channel) relate to agents of near transfer of solid material (deposits, the weighed particles, etc.), and the 3rd type to agents of distant transfer, but also by where hydrodynamic zones the geosystem is spatially placed—in a zone of drain formation, in a transit zone or in a zone of drain carrying out, in a zone of dispersion or accumulation. The last one depends on a relief that defines the intensity or absence of linear erosion.

River basin of the Ertis created in Altai mountains is a product of all interaction factors, forming macrogeosystem, and depending on conditions of development of the latter, the "hydrological" (erosion normal) link of physiographic processes has to change as well. Modern backbone processes of the basin are closely linked with developments of the water currents, making the top links of hydraulic network.

All phenomena and processes, connected with the interactions of the streaming water and spreading rocks, represent united erosive and accumulative process. In this regard in basins of the above-mentioned rivers it is possible to separate geosystems with the dominating non-channel streams, formed under the influence of waters, flowing down from the slopes, making plane erosion and not concentrated accumulation. Other geosystems function in the conditions of plane erosion. Parameters of water and thermal balance are steadier here, and that affects also stabilization of bioproductivity.

Temporary channel streams, carrying out the linear (ravine) erosion and its development represent a self-excited process and concentrated accumulation in the form of carrying out cones characteristic for many basin territories. Especially they are distinctive for zone of drain dispersion concerning the small river basin and low-hill terrain regions. Upper courses of all rivers have the constant channel streams, formed during geological history with well expressed river valleys constantly punching the forms of channel relief created by them.

Accumulative processes are distinctive for lower reaches of the basin, and also for zones of drain dispersion. Erosive processes dominate in the Uba and Ulba river basins. Generally, erosive and accumulative process is the only one in many river basins of the III and IV order playing the predominating role in the development and dynamics of geosystem.

Water streams are created under certain conditions; on which combination of water content size, mode, and longitudinal bias of courses, formed structure of alluvial rivers are depended. All this defines erosion intensity on slopes and channel forming activity of constant streams. Erosive-accumulative and channel processes are the unified whole, depending on the nature of reservoir geosystems and it is also impossible to consider them separately from the river. Therefore, also their reservoirs have to be studied in close interrelation and interconditionality.

Tectonic and climatic features influence unequally a fluvial denudation of gecosystems' zones drain formation, therefore depending on a combination of a change tendency concerning these factors, opposite relief-forming processes can develop at the same time within the same river basin. Geosystems of zones drain formation concerning the Altai main rivers function in the conditions of different genesis temporary channel streams and differently react to changes of climatic elements.

Facial heterogeneity of the geosystems concerning zone of drain dispersion is connected with the formation of bald mountains, swamps, hills and ravines.

Erosive and accumulative activity of the Altai rivers is connected with expenses of energy stream between which ratios is approximately defined by the work of water consumption on a bias and, therefore, equally reacts to fluctuations of climate, leading to change of water content stream, degree of drain unevenness, etc., and the tectonic movements directly causing reduction or increase in a longitudinal bias. The water movement is connected with energy transformation and ability of continuous change of the hydraulic stream characteristic. Therefore, the degree of geosystems stability in different zones of drain is various, and depends on the aforesaid.

Processes of deep erosion or accumulation depending on the factors causing them, extend down (transgressively) or up (regressively) on a stream, and also can be shown at all river length. Erosion depth gives also intensive side erosion, which can be shown also in an accumulation zone on slopes. The degree of geosystems variability of the Ertis river basin, characteristics of transformational ranks of biota and soil depends on deep erosion intensity.

In the inundated geosystems in connection with the abundance of the latest years' water and global warming of climate, the erosion basis has been changing. Increase in erosion basis owing to anthropogenous factors and increase in a glacial drain have increased the level of ground waters in recent years, and that was reflected in a bioproductivity of geosystems flooded territories, in haymaking and postural grounds, the above flood-plain terraces, which are formed on alluvial-proluvial deposits.

The intensity of above-named processes caused by anthropogenous factor creates a peculiar complexity, mosaicity of inundated and valley geosystems of the Ertis river basin.

2.1.2 Principles of Identification and Differentiation of Geosystems in East Kazakhstan

In the systemic researches of basin territories there appeared two directions: planetary (regional) and typological (intrasystem). Regional and typological geosystems function in conditions where radical and lateral links of substance streams have identical value. Links of these streams are different, unstable in macrogeosystem time, i.e. stability of equilibrium state is various. The researcher needs to consider geosystem not only as link of natural components. It is proved that this connection becomes interrelation, only in case interaction function begins. This scientific direction is developed in V.B. Sochava's works.

The second direction recognizes only lateral (gravitational) components of connections directed from a watershed to basis and along erosive forms of relief. The essential moment of developing gravigen geosystems is the tension and capacity of gravitational streams at the corresponding stages of substance circulation (Solntsev 1975).

From the systemic positions both approaches are correct, their equivalence is caused by "polydegree of structure" according to V.N. Solntsev and "polysystemacity" according to G.S. Makunina (Makunina 1980, 1983).

Geosystems of the Ertis river basin are paragenetic natural and territorial complexes, combined by unity of vertical and horizontal currents of substance and energy, i.e. mass and energy exchange and also formed in the conditions of one litogenesis and one direction of geographical drain.

A.D. Armand determines geosystem as "any set of the interacting elements". The system is capable to be divided into subsystems and, in turn, can be included in the highest order.

All river basins of the Ertis are defined by us as united macrogeosystem, which structures are intergeosystems and subgeosystems. Existential connections of channel formative processes are essential in mesogeosystems development.

Functional integrity of each subgeosystem is defined by development of vertical material and energetic circulations under the influence of atmosphere energy and biota activity, imposing of small substance circulations on gravigen streams and partial involvement of the latter in biological and atmospheric migration and energy. Each subgeosystem is presented by river basins—inflows of the III-IV order, the so-called small rivers (Uba, Ulba, etc.).

The river basin is the typical gravigen geosystem—orographically limited, functioning as a united natural body in which division of matter is caused by realization of gravity energy put in it. Potential size of energy weight is defined by difference of absolute heights and the types of relief (topography).

Dynamic activity of gravigen geosystems is shown in intensity of material and energetic streams, determined by surface biases, their lithology and bioclimatic situation. Bioclimatic conditions together with amplitude of alignment surfaces or absolute altitudes define higher dynamic activity in the gravigen subgeosystems with more steep slopes, pliable to aeration and demolition by breeds. In subgeosystems, functioning in different bioclimatic conditions, higher dynamic activity is inherent in the increased moistening subgeosystems.

So, at one hypsometric level functioning of the Kyzulsy-Taintinsky and Sharsky subgeosystems differ from each other visibly. The first is presented by the dense vegetation, which is a mechanical barrier on the way to transit substances, their potential forming a biogenous barrier, strengthening material energy potential of all subgeosystem. Capacity of gravitational streams in Sharsky subgeosystem depends on amount of precipitation, lithologic properties of breeds.

Thus, the hierarchy of gravigen geosystems has been defined. It is the largest unit of all subgeosystems main inflows, orographically limited by space, accurately delineating gravigen geosystems of III and IV inflow orders (Fig. 2.1).

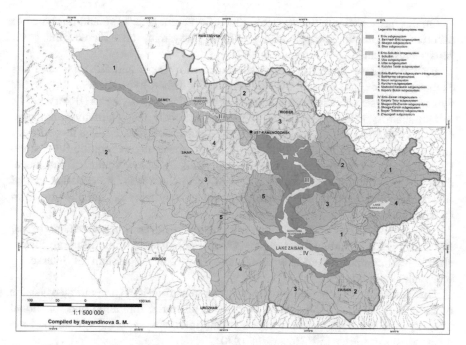

Fig. 2.1 Map of the Geosystems in East Kazakhstan

Within the last ones, gravigen systems of slopes are functionally complete, which are complicated linearly-cascade gravigen systems of ravines, temporary and direct water currents. Linearly-cascade gravigen systems are the integrated geosystems, symmetrically and consistently uniting material and energy stream crossed by gravigen geosystems of slopes and small river basins.

Structural connection of the gravigen geosystems are geosystems of the first topological level type, in which tension and capacity of through and "gravitational" streams is the power and migration activity of the substance. However, the linearly—cascade gravigen system can not be a part of the gravigen geosystems, and take an autonomous position in relation to them (for example, a plakor), and in this case geosystems of slopes and erosive forms are "open" for the intake of substances and energy from watersheds of the first type geosystems.

F.K. Milkov called geosystems with the unidirectional streams of connections in their borders—par dynamic, and the geocomplexes composing them—par genetic, i.e. par genetic complexes are characteristic for all geosystems, development and functioning of which is referred to a superficial drain. But when geosystems are formed on different breeds, and sometimes in the conditions of different natural zones, definition par genetic is not appropriate, since it contradicts a concept of par agenesis (Milkov 1972, 1981).

On this basis, structural geosystems parts of different genesis should be related to different par genetic groups. Such gravigen geosystems also include the basin geosystems created by river sources.

By the type of intensity of the substance movement and their power mass exchange and degree of stability, we have distinguished hypo dynamic and hyper dynamic geosystems, being compounds of sub—and intergeosystems (Bokov 1977).

Domination of fading processes is the characteristic feature of hypo dynamic macrogeosystems with the weak movement of substance streams. The insignificant anthropogenous factor can quickly bring it out of an equilibrium state. Hypodynamic microgeosystems are characterized by heavy traffic of substance, their power activity quickly leaves a natural cycle since it is in a constant exit of substance, power tension is weak or absent. Such geosystems can quickly transform to other genetic row.

The general interconnecting signs are characteristic for all geosystems of the Ertis river basin: continual and discretised in the existential-time relation; hierarchical; multicomponent and dynamic; also they are the system of interacting morphological parts.

Distinguishing of connection types between geosubsystems and intergeosystems of basin defines their geographical unity with territorial limitation and functional integrity inherent in them. Intercomponent and intergeosystem links in all the Ertis macrogeosystem refer to it (Ioganson 1970).

Thus, large taxonomical units of all Ertis macrogeosystem are revealed as united geosystems with the universally oriented substance movement in the area of drain from a watershed to the basis of erosion, and are defined as sub-intrageosystem. On zone of drain tanzit in valley of the Ertis river we have revealed four subgeosystems: Ertis, Ertis-Shulbin, Ertis-Bukhtyrma, Kara-Ertis-Zhaysan (Bayandinova 2005a, b; Dzhanaleeva 1998) (Fig. 2.1) (Bayandinova 2003a–e).

In the Ertis macrogeosystem, created by geosystems, connected with the basin of the Ob river large inflow, taking into account the typification of landscapes, also seventeen subgeosystems have been distinguished, each of which has been formed and is functioning in the conditions of the III-IV order river inflows relatively to the main course.

Each of them is divided into four zones linked to superficial drain: geosystems of drain formation zone; geosystems of drain transit zone; geosystems of drain feathering-out zone; geosystems of drain dispersion zone. Geosystems of drain formation include the geosystems united by lines of highlands and represent intensive compartmentalization massif of the Alpine shape. They are put by granites, gabbro, gradiorite, effuziva and their tufa, effusive and sedimentary breeds, rare sandstones, slates, marble, and focused almost in the width direction according to general direction of geological structures and zones of large technical violations. Ridges have received a modern outline as a result of tectonic movements, which were more active at the end of Paleogene and have been continuing so far.

General character of highlands compartmentalization of all Ertis macrogeosystem extends also to backbone processes.

High position of glaciers defines the most complicated structure of geosystems, unlike the basins of those rivers, where freezing marks at the lowest. The development of geosystems happens in the conditions of semi-humidified climate here.

Geosystems of transit zone are related to the middle and low-mountainous terrain and are formed in the conditions of terraces, leveled terrace shaped surfaces on slopes and in the bottoms of mountain hollows with traces of ancient freezing. The strengthened erosive activity with expressed alternation of the glacial accumulation periods of deposits with the periods of erosive activity strengthening in rivers are characteristic for river basins of the Shar, Ulba, and etc.

The genetic unity of such geosystems is obvious. The main characteristic processes, influencing backbone is colluvial accumulation, soliftual processes, the expressed benching, slope accumulation. Morphological structure of gecosystems on such sites is also complex and signified by transition state from mountain-steppe upland geosystems to shrubby-rich in cereals with forest geosystems, and to an ephemeral rich in cereals—shrubby, formed along the course and in a high flood plain.

The geosystems of drain feathering out zones are difficult for studying in connection with strong anthropogenous changes.

The powerful layer of alluvial-proluvial deposits, their joint with the material of terrace above flood-plain creates difficult character of soil formed breeds concerning carrying out cones.

In the friable deposits of carrying out cones the level of ground waters are high, some the mountain river courses are feathering out again in that zone.

The geosystems of drain dispersion zone function in the conditions of spread-eagle drain, develop in the conditions of insignificant biases. Development of geosystems happens in the conditions of insufficient moistening.

Thus, in connection with the aforesaid, a major factor of the Ertis river basin geosystems organization is the nature of interrelations of litogenesis and superficial drain, as well as the uniting properties of gravigen, which determines their stability and consistency in space and time.

Huge scales of industrial and agricultural production of the Ertis macro-geosystem strengthen a negative ecological situation. Annually more than 250 million m^3 of sewage is dumped in the river Ertis. Pollution of these waters on the weighed toxic substances exceeds by 3, 6 times as to maximum allowable concentration. Dumpings from the city treatment facilities of Ust-Kamenogorsk, Semey are the main source of the Ertis river pollution (Bayandinova 2005a, b). Heterogeneity of modern technical equipment in the conditions of market economy, widely extended injurious, irrational use of natural resources, led to the adverse natural and anthropogenous processes in big areas. All this has brought to the phenomenon which has received the name of "ecological crisis" in the scientific literature. The problem of geosystems protection from negative processes of techno genesis became one of the major urgent practical and natural—scientific problems in the Republic of Kazakhstan, where geosystems of East Kazakhstan demand a detailed geoecological assessment.

2.1.3 Characteristic of Geosystems

The Ertis macrogeosystem which is compound of Karsko-Ob megageosystem in territory of the Republic of Kazakhstan is presented by four subgeosystems, which unite basins territories of this river numerous inflows, constant or temporary drain of which is directed towards the Ob River. The Ertis, Ertis-Shulbinsky, Ertis-Buktyrminsky, Ertis-Zhaysansky are related to them.

Northeast watersheds of the Ertis-Buktyrminsky subgeosystem are presented by the Koksuysky ridge and Listvyaga, which morph structure is close to the Alpine lines. The Ertis, Uba, Ulba inflows originate from the southern slopes of the Koksuysky ridge, which forms Actually-Ertis subgeosystem. The rivers Buktyrma and Naryn which basins form Ertis-Buktyrminsky subgeosystem flow along the southern slopes of the Listvyaga and Ulbinsky ridges.

The Kurshym inflow originates from the southern slopes of the Naryn ridge (absolute height is 3375 m). The river Kaldzhir follows from the mountain lak which is the right inflow of Kara Ertis, and its drain is regulated. Basins of these three rivers create Buktyrminsky, Ubinsky and Kurshymsky subgeosystems, and their functioning depends on physiographic processes of all Ertis macrogeosystem (Bayandinova 2005a, b; Dzhanaleeva and Bayandinova 2003).

The watersheds of the Ertis macrogeosystem southern suburbs are occupied with the massif of Tarbagatai ridge, which has the insignificant amplitudes of geotectonic movements. Southwest watersheds are provided by Shyngystau low-hill terrain (Edrey, Arkat, Murdzhik, etc.). Northeast part of intrageosystem is occupied with suburbs of the West Siberian lowland—the Kulunda Steppe.

In tectonic relation the southern part of the Ertis macrogeosystem represents uplifted pool of the neogene-downfourth raising, formed on the place of her cynic constructions. In Pleistocene these territories have undergone a freezing. Northern suburbs of the region's macrogeosystem belong to the West Siberian plate.

The mountains of strongly dissected relief of southern region suburbs gradually passes into wavy and hilly plains of the Ertis average current. Absolute marks vary from 235 (firth of the river Uby) up to 2000 m (at tops of "snow covered mountain peak").

The surface of the Ertis macrogeosystem, hollow inclined to the north, has very difficult relief. Often flat manes alternate with gentle slopes, sometimes with the small closed lowlands located in chains. The largest hollows are connected with tectonic motions (Lake Teniz).

Neogene deposits are presented by two suites—the lower Miocene (the Aral suite) and average Miocene (the Pavlodar suite). The first one is of lake origin—the salted green clays with plaster, the second consists of lake-marsh and alluvial rainfall of red color with plaster. Quarternary deposits are widely presented, their power depending on local conditions of accumulation and destruction vary from 0 to 100 m. Most often we can meet sandy, alluvial deposits, lake and marsh accumulation of different mechanical structure and integumentary yellow-brown carbonate loams of forest shape.

Underground waters are fresh. Lime stones of kembro-silur are the most water-laden, where waters are of fissure-karst type and rigid. Detrital glacial and alluvial deposits are also water-bearing. Reservoir waters in alluvial columns of ancient river valleys are characteristic. Waters have weak chloride-sodium salinization.

The most various soil climatic conditions are characteristic for the Ertis macrogeosystem. In the Ertis-Zhaysansky subgeosystem with very dry north wind and domination of ephemeral desert vegetation, the appraising points of soil climatic conditions are less than 40. Mountain territories of east suburb subgeosystems concerning macrogeosystem belong to a damp mountain agroclimatic zone and are estimated at 100–130 points. Further to the north they decrease to 60–80 points.

Except the listed inflows, in physiographic situation of subgeosystems the rivers like Shar, Shagan, Ashisy, Kyzylsy falling into the Ertis plays major role. Some of them have the outlined above flood-plain terraces, but the drain in them is changeable (Shagan, Ashisy, and etc.).

Rivers due to their mode refer to the Altai type. Nourishment of the rivers is mixed due to melting of seasonal snow and summer rains, apart from the Buktyrma river having glacial nourishment. On the rivers Kurshym, Buktyrma timber-rafting is carried out. All rivers (except for the Ertis) have hydro carbonate composition of waters in the period of flood and chloride structure during the low-flow period. In low-flow period there is an increase in mineralization by 2–5 times. The current freezing is developed in the Katun range, in the sources of the rivers Bereli and Sarymsakty.

Average annual layer of drain fluctuates from 1000 to 1500 mm (in a zone of drain formation concerning macrogeosystems) and decreases to 2–5 mm in a transit zone.

Average long-term consumption of water makes 895 m^3 per hour in the reservoir area of 179 thousand km^2 Average annual water supply of the Ertis macrogeosystem, formed within the Republic of Kazakhstan, makes 200,000 km^3 on 1 km^2 in a zone of drain formation.

Flooding type is spring and summer. The beginning of flooding is from the 10th of April till the 31st of March in a zone of drain formation and on the 5–10th of April in a transit zone. The ending of flooding is on 31st of July 31 and 15th of May respectively.

The main waterway of Kalgaty-Takyr and Shorga-Kostin subgeosystems is the Kara Ertis river, which has a well-developed valley. At a confluence with lake Zhaysan it forms the boggy delta. Lake Zhaysan occupies the ancient Ertis valley. Lake waters are fresh and flowing. After creation of the Buktyrminsky hydroelectric power station and reservoir, the waters sub time on the top of relief was extended on the Ertis valley to Zhaysan, and level of the lake has increased to a mark of 388 m, as a result the low coast and the delta of Kara Ertis have partially been flooded. The river Kaldzhir, Kurshym, Kendyrlik, and etc. flow down from slopes of nearby mountains. Many of them dry up in the low-water period.

The Ertis-Zhaysansky subgeosystem is formed by large lake—Zhaysan, being young geological formation. Lake Markakol is located at the height of 1449.3 m above the sea level. Markakol-Karakabin subgeosystems function in the conditions of strongly dissected relief. The slopes of the middle mountains, turned towards the water lake area are occupied with mountain-tundra, mountain-forest and mountain-meadow-steppe natural complexes. The geosystems function in the conditions of increased moistening. The geosystems, created by 27 small rivers (Topolevka, Karabulak, Matabai, and etc.), have steady character. Ultrafresh, subacidic water of lakeside territories forms geosystems of calcic group. The Markakol national nature park is organized in 1976 for protection of biota (Bayandinova 2003c–e).

Geosystems at the height of 1760 m above the sea level are created by lakeside geosystems of Lake Rakhman, surrounded with massif forests from larch, cedar and fir-tree. The lake is flowing and has depth of 30 m. Average annual fluctuation of water level makes 1.5 m.

The lowest rank geosystems are formed in a zone of drain formation in the conditions of mountain-tundra, mountain-meadow, mountain-forest, mountain-steppe high-rise belts. Woods from the Siberian fir and larch are characteristic for the northern massif of macrogeosystem.

The geosystems connected with drain transit zone, develop in the conditions of steppe zones. Geosystems of flat plains, with numerous suffusion and relic thermokarst kettle and drain hollows dominate, that causes weak fitness and complex combination of bogging and salinization processes. Instability of moistening, its intra annual fluctuations lead to alternate strengthening of one or other processes.

Not salted soils prevail in the Ertis-Buktyrminsky subgeosystem. The saline soils can be found on the coasts of salty lakes, and the meadow types are in valleys of some rivers. General direction of geochemical drain in this subgeosystem from the south to north, and the number of salted soils in this direction gradually decreases. Such situation is explained by the change of moistening coefficient, first of all due to the reduction of evaporability. This phenomenon is called inversion of salt belts. The main type of soils salinization is sulfate-sodium. The tendency of the salted soils increase in the area doesn't pass a toxic threshold.

The big contrast, connected with diversity of habitat communities is characteristic for biological circulation and production processes in biota of the Ertis macrogeosystem. According to long-term stationary researches in the Barabinsk forest-steppe, stocks of live meadow steppe phytomass on ordinary black earth (tops and the top slopes of manes) make 16.4 c/ha (including 2.2 c/ha an elevated part), annual efficiency—19.0 c/ha (including 4.0 c/ha of elevated weight). The maximum efficiency is noted for inundated reed grass swamps (63.7 c/ha), minimum for sea blite thickets on meadow saline soils (3.1 c/ha). Birch splitting on manes product 9 c/ha (7 c/ha of elevated parts) of phytoweight, and in the inter low ridge reduction—13.8 c/ha.

Stocks of cindery elements and nitrogen in live and dead organic mass of Ertis macrogeosystem make 570 kg/ha in meadow and saline communities, 1600 kg/ha in meadow steppes, 9200 kg/ha—in reed grass swamps. In underground bodies about 80% of mineral elements are concentrated. In order to create a year production in the meadow steppe 1013 kg/ha of cindery elements are consumed (Ca, Na, K, Si), 175 kg/ha of nitrogen, in a birch splitting of inter low ridge decrease are much less (in the sum to 454 kg/ha). High intensity of metabolism is characteristic for the meadow steppe, and the biological circulation is almost closed.

Soil cover in hollows and lowlands is variegated. Leached blackearth prevail in the north of the Ertis macrogeosystem on watersheds under mellow meadows, semi terrestrial meadow and black earth soils are widespread in the south under meadow steppes. Various halophytic options of meadow steppes are widespread along river courses, on terraces above flood plain. Gray forest solodic soil is developed under birch splitting on manes, and on the kettle is malt. Terraces above flood plain of all subgeosystems are presented by low alluvial clay and loamy plains, and also ancient lake and alluvial with drain hollows. Halophytic options of steppes— fescue-feather grass with halophytic forbs and fescue-goldilocks on sodic soil are widespread in low terraces, hollows, lake kettles of the Markakol-Karakabinsky subgeosystem and the Ertis-Zhaysan intrageosystems. Low alluvial and Aeolian sandy plains along sandy terraces above flood-plain and ancient deltas are characteristic for Bugaz-Tebestinsky subgeosystem, and it is frequent with dune-hilly and hilly-grown in beds sands, semi-fixed groups from sandy feather grass, sheep fescue, and fatuoid and psammophytic forbs on undeveloped dark-chestnut and chestnut soils.

The Ertis-Shulbinsky, Ertis-Buktyrminsky, Ertis-Zhaysansky intrageosystems cover geosystems, which are related to the valley Kara Ertis, lake Zhaysan and the Buktyrminsky reservoir. They have typically Central Asian semidesertic lines. The territories with the lowest marks are presented by the steeply-sloping plain with tasbiyurgun-wormwood vegetation on brown soils. The valley-growing geosystems of Kara Ertis: Kalgaty-Takyr, Shagan-Ob-Zharmin, Shorga-Kostin, Bugaz-Tebestin, Zhuzagash have salino-wormwood vegetable communities, and lakeside winnow anew sandy massifs are Erkek-takyr. The high terraces of Kalgaty-Takyr subgeosystem are occupied with wormwood forbs associations created on light-chestnut soils. Above, according to high-rise belts of Shorga-Kostin subgeosystem change the shape from mountain-tundra, mountain-meadow to mountain-forest and mountain-steppe.

Natural complexes of Ertis-Zhaysan intrageosystem markedly differ from each other with modern physiographic processes that are caused by various conditions of all geographical drain formation, the cornerstone of which is superficial and underground drain.

Factors of techno genesis in different degree changed the natural capacity of region and its ecological situation. Subgeosystems of the Bukhtyrma, Kurshym, Kalgaty-Takyr develop under the influence of toxic substances, which are products of the disintegration emissions from the enterprises of nonferrous metallurgy. The water and land resources, the air basin of city agglomerations of Ust-Kamenogorsk, Zyryanovsk, Ridder, Serebryansk are polluted by salts of zinc, lead, mercury, beryllium. Especially, high pollution by lead (3.3 maximum allowable concentrations) should be noted.

The huge scales of industrial and agricultural production of the Ertis macrogeosystem strengthen a negative ecological situation. Heterogeneity of modern technical equipment in the conditions of market economy, widely extended injurious, irrational use of natural resources led to manifestation of adverse natural and anthropogenous processes on big areas. All this has brought to phenomena, which is named as "ecological crisis" in scientific literature. The problem of geosystems protection against negative processes of technogenesis became one of the major practical and natural—scientific tasks in the Republic of Kazakhstan. However, at the same time there is a certain contradiction between the public nature of conservation and private activity of many enterprises of nonferrous metallurgy that brings certain difficulties in the solution of environmental problems of the region. In 1993 the Ministry of environmental protection gave the East Kazakhstan region the status of a zone of ecological catastrophe. However, at the same time there is a certain contradiction between the public character of conservation and private activity of many nonferrous metallurgy enterprises that brings certain difficulties in the solution of region's environmental problems. The East Kazakhstan region was given the status of an ecological catastrophe zone in 1993 by the Ministry of environmental protection.

According to the Federal State Statistics Service, only in 1996 year 375 cases of toxic substance emissions in the rivers Tikhaya, Krasnoyarka, Beksu, Ulbu, Glubochanka were recorded. And the threshold limit values in some cases almost hundred times were higher than normal. Only to the river Beksu 160 emergency emissions have been recorded. To the river Glubochanka, the zinc emissions exceeding maximum allowable concentration by 240 times are recorded (Bayandinova 2003b–e).

The Actually-Ertis subgeosystem occupies the territories related to a drain of the river Ertis at the exit from the mountainous territory. Absolute marks of the valley don't exceed 200 m. Natural complexes of Shar and Shagan subgeosystems are created on the left bank, and has general inclination on the northeast to the river Ertis, also breaks a low ledge. The relief is presented by the flat plain with small ridges and ravines. Some inflows fall into the Ertis, but many of them come to an end in drainless small lakes. Small lakes (Shureksor, and etc.) are located as two parallel chains along the modern course of the Ertis and represent a bottom of ancient valley. The relief of geosystems of the Ertis left bank is poorly dissected, and neogene alluvial deposits are blocked by loess blanket at the power of 10–20 m everywhere. On sublime sites the power of loess makes 40 m. The powerful cover of the loess provides good aeration and drainage, in this connection soils are united and almost completely opened here. Some lake hollows are cut in 70–80 m (lake Kyzylkan, and etc.).

Geosystems of the right bank are formed in the conditions of overlapping relief. Height of ledges reaches 12-15 m. Natural complexes of Balapan-Ertis subgeosystem are created in the conditions of rolling and flat alluvial plains of the fourth terrace concerning dry lake hollows and hills. Modern small lakes with mineralized water (Small and Big Yamyshevsky, Tobolzhansky, Karasuysky, Prigonnoye, and etc.). Geosystems of Shar and Shagan subgeosystems, are created on the most ancient plain with strongly dissected relief and occupied with dune massifs. On the Ertis's right bank neogene deposits are blocked by ancient the sandy alluvial of this river, having power on average of 5–10 m therefore soils here easily flutter. To the east, they have heavier mechanical structure.

Geosystems related to the first, often saline, and second terrace of the Ertis towering on 4–6 and 15–17 m over the river, are created on both of its coast in the form of strips with a maximum width up to 25 km. Geosystems of the third terrace stretch at the height of 28–32 m over the river, and are widespread basically on the left bank and reach the Kazakh hillocky area. They are mainly composed by pebble-gravel material, gradually changing into sandy to the North.

Geosystems functioning of the lowest order is due to the following features: saline and meadow-chestnut soils with complex fescue-feather grass steppe and halophytic meadows are developed on a left bank in hollows; light loamy, sandy loam dark-chestnut soils with fescue-feather grass and tyrsovy-feather grass vegetation on loams with psammophytic forbs in easy soils are developed on flat plains. Geosystems function in the conditions of kettle-flat moderate and dry steppe.

Geosystems of the Ertis valley in the central parts function in the conditions of typical chestnut soils and are created on the left bank on the second and third terrace, composed by low-power sand-pebble river deposits with fescue-tyrsovy steppes and wormwood-fescue communities on sodic soils. Halophytic meadows are developed in low lands. Pine banded forest with pea tree grow on ancient hollows of drain and sandy deltas of the dried inflows.

Geosystem terraces of right banks are put by sands, and degree of their winnow anew depends on age of terraces. The southwest suburb of Kulunda steppe is complicated by strips of the well remained Ertis crease, hollows of salty lakes up to 60 m in depth, and massifs of dunes and loessial ridges (Balapang, and etc.). Vegetable communities are presented by sand-feather grass steppes with spirea thickets. Pineries with light forests grow on the southern sandy massifs (North Srostensky, Sosnovsky, Chaldaysky pine forests, and etc.). Natural complexes functioning of the Actually-Ertis's subgeosystem part happens in the conditions of the dry steppe. Geosystems are used as high-yielding water meadows. After the construction of the Ust-Kamenogorsk, Buktyrminsky dams and hydroelectric power station, flooding of flood plain is stopped, and it lost the biological value.

Besides the erosive processes, accelerating negative consequences of techno genesis, negative influence of nonferrous and ferrous metallurgy branches and in general the mining industry on hydro chemical mode of geosystems, on all natural capacity of the region should be noted.

In the seventies, scientists of Academy of Sciences of the Republic of Kazakhstan developed series of actions aimed at improving the Ertis river basin for agricultural use. Intensive deflation on all depth of humic horizon, an exposure of strong carbonate breeds has caused increase in inarable land for agriculture. Now land grounds are often used as pasturable and haying grounds. The meadow areas have decreased after the construction of Buktyrminsky hydroelectric power station. Acres were reduced from 78 to 17%, however pasturable and haying grounds have increased from 20 to 72%. Again created anthropogenous modifications of pasture and haymaking landscapes, and also agro landscapes now experience intensive technogenic pollution. In recent years, natural capacity of the Ertis river basin geosystems has sharply decreased in connection with the influence of the enterprise's harmful toxic emissions concerning mining, color, chemical industry and power. Industry objects have changed the natural environment of geosystems functioning, and it negatively influences its physical and geochemical compound. In general, all these processes have changed the general course of mass and energy exchange. Besides, the influence of techno genesis negative processes is aggravated by jubate-hollow character of relief, halophytic biota, and also activization of air streams pollution from industrial complexes. Except efficiency decrease in connection with pollution, these soils easily give into deflation, as they are presented by ancient easy sandy alluvials. Geosystems of the Ertis river left bank have good conditions for a drainage, however influence of the basic northwest and western transfer is smoothed by high coefficients of pollution.

Annually, more than 250 million m^3 of sewage is dumped into the Ertis river. Pollution of these waters on the weighed toxic substances exceeds by 3, 6 times of maximum allowable concentration. Dumpings from city treatment facilities of Ust-Kamenogorsk, Semey are the main sources of the Ertis river pollution (Bayandinova 2003a).

The population density of the Ertis macrogeosystem is high. The population density along the river valley and in the industrial centers of East Kazakhstan is especially high. In these regard, significant areas of intended for building, industrial modifications of landscapes and agro landscapes are developed. For the functioning of such geosystems more than 3 billion m^3 of water is got from an underground and superficial drain, and more than a half of it is dumped into the Ertis river, making negative impacts on ionic structure of drain and saturating it with toxic ingredients. Besides, the sources of natural waters pollution are the atmospheric precipitation, bearing a large amount of polluting ingredients of industrial origin which are washed away from air. At running off on slopes atmospheric and thawed snow are in addition washed away from the surface of soil pollutant. In city conditions water flows bear a large amount of oil products, acids, and highly toxic elements. The city sewage contains various products of disintegration of radioactive thermoelectric origin, caused by the influence of the Semipalatinsk nuclear test sites (Cherednichenko 2002; Garmashova and Cherednichenko 2002).

In order to restore natural potential, there is a need to rezone areas, occupied with dry-land agriculture on irrigated ones, because the grain farm is unprofitable here and demands not only huge financial expenses, but also the solution of manpower problems caused by the population migration. Problem solution of water resources use has to be solved towards maximum reduction of dumping polluted river waters, ground water conservation in rolling interfluvial territories, purifications of industrial waters, introductions of water supply turnover in industrial centers.

Four subgeosystems are included in the Ertis-Shchulbinsky intrageosystem (Shulbinsky, Ubinsky, Ulbinsky, Kyzylsu-Tantisky). Each of these subgeosystems has united factors of technogenic pollution and identical conditions of environmental risk. It is connected with the influence of atmospheric emissions, waste and industrial drains. Nevertheless physiographic conditions, which gain local lines, cause various degrees of negative ecological situations: these issues in more detail will be considered in the following subsection.

2.2   Analysis of Major Factors in Formation and Migration of Technogenic Polluted Geosystems in East Kazakhstan

2.2.1   Principles in Studying and Mapping of Technogenic Geosystems

The qualitative structure of natural migration flow changes as a result of economic activity (because of technogenic substances (TG) inclusion and energy while forming technogenic geosystems).

In the scientific literature much attention is paid to the questions of lands' damage in connection with the production and processing of minerals in the zones of nonferrous metallurgy impact, and also chemicals, introduced to the elements environment. The share of the research, covering at the same time all natural components as the soil, plants, water, and air is very small. Complex predictive landscape and geochemical researches were conducted under the leadership of M.A. Glazova, who analysed the influence of industrial centers with various types of technogenic impacts on geosystems in geochemical relation. But such works are not sufficient, and they, as a rule, cover small territories that limit possibilities of geographical interpolation of the obtained data (Panin 2000; Glazovskaya 1962, 1967, 1976, 1992; Geohimiya tyazhelyih metallov v prirodnyih i tehnogennyih landshaftah 1983; Glazovskaya and Kasimov 1987).

Insufficiency of methodical and methodological armament of landscape and geochemical researches, at complexity and diversity of problem, forces to draw a close attention, first of all to the principles and technique of these researches. Methods, which give satisfactory results on traditional landscape and geochemical researches aren't enough for the analysis of territories, undergoing technogenic factors influence. It is concerned with the system of field and analytical methods as well (Dyakonov 1985, 1988).

Studying landscape and geochemical features of environment is impossible without the simultaneous solution of two interconnected question groups: a) the main types identification of local responses concerning natural systems on technogenic influences, i.e. analysis of structural and geochemical reorganization of initial landscape and geochemical systems; b) distinguishing geosystems with the united type of responses to certain technogenic influences, i.e. division into districts of territories according to groups and classes of technobiogeom (Glazovskaya 1988; Kosimov 1980; Aslanikashvili and Saushkin (1975); Perelman 1966, 1968). In other words, it is necessary to combine landscape forwarding works, detailed researches and the wide geographical and landscape and geochemical analysis of the environment with the generalization of available field materials.

Basic continuity of small-scale and large-scale landscape and geochemical researches defines the main methodical scheme of researches—the principle of consecutive landscape and geochemical analysis (Nikolaev 1978, 1989).

Without getting into the classification analysis of technogenic transformed landscapes, proposed by different authors, we will mark that in most of them technogenic factors are accepted as a leading change of the environment, and they can be quite reasonably compared with the intensity of geological influences.

Actually, structure and qualities of technogenic transformed territories is a result of hyper gene processing concerning initial substance reserves, and those which come to them at migration. Active withdrawal of substance big mass or essential increase in substance or energy receipts in natural systems is followed by changes of geochemical qualities of the last, and intensity of such transformation depends on that as how many arrives or withdrawn from natural circulations. Therefore, at researches of geochemical features, technogenic factors have to be considered along with natural ones. Thereby, geochemical activity of the prevailing technogenic influence types of natural system responses also have to be investigated.

In the case of mining production and the related mining and processing industry the following have to be studied: (a) extraction from a subsoil minerals, their transportation and the primary TG dispersion of substances of various degree of geochemical activity and toxicity, caused by it; (b) processing, i.e. enrichment of ores and receiving concentrates, storing of waste and secondary technogenic dispersion of substances.

Geosystems in mining areas also change in connection with the intake of flotoreagents, power production waste in natural circulations.

In addition, there take place mechanical violations of the initial geosystem because of the withdrawal or damages of soil, ground or organic masses. Mechanical influences cause changes in the mode of migratory processes that also leads to geochemical environment reorganization, but in another direction and intensity.

Therefore, while choosing reference objects, first of all, it is necessary to consider the geochemical activity of the operating TG loadings, the degree of concentration or substance dispersion in the course of economic activity, and also their natural prevalence.

Taking into account these factors, the influence of TG in connection with the intake or withdrawal of elements with high (Si, C, S, Fe, etc.) or low (Pb, Zn, Sn, etc.) and very low (Hg, Au, U, As, Cd, etc.) natural percentage abundance has to be isolated. The concentrated impact on natural processes of elements connection with low and very low percentage abundance is mainly technogenic, since in the natural state they are characterized by very high extent of dispersion (Dergunov et al. 1988).

Geochemical active substances possess, as a rule, high biological activity. With the intake in quantities above critical level (miscellaneous for different environment) there appear tangible deviations in biological systems from the norm. Among geochemical active elements and connections, especially with low and very low natural percentage abundance, there are a lot of highly toxic ones. It is natural that geochemical stability—adaptation or reorganization of natural systems geochemical structure or their components to the different TG groups of influences isn't identical. It is necessary to differentiate not less than two cases:

1. technogenic influences are toxic for the environment, they cause its degradation, breakdown or radical reorganization, up to destruction of natural geosystems or their separate components;
2. technogenic influences are nontoxic for the environment. Such impact can even be favourable (intake of minerals, insufficient for specific landscape and geochemical conditions, lime application or plastering of soils, introduction of organic fertilizers, etc.). However, technogenic influences, though being nontoxic can be unfavourable for the normal functioning of natural systems.

In mining areas, because of the distinctions in the TG influences' dynamics, two types of the environment transformation are combined and alternate spatially.

1. At continuously operating or periodically renewable TG loadings in the areas of mining functioning and the mining processing enterprises, central heating and power plant, hydraulic engineering constructions, and etc. degrade and collapse initial natural communications, and the system of geosystems transformation zone is formed, which is characteristic for each TG loading type.
2. After working off the field stocks and termination of mining operations, the broken natural systems or their separate components are gradually restored, however such restoration is not always possible without the application of remediation actions. Peculiar zones of restoration with geochemical qualities are formed, significantly different from the initials. But along with the restoration of soil and geochemical interfaces in such areas there is also a further degradation of some territory sites under the influence of residual technogenic factors—breeds dump of different degree on toxicity, shoddy hydraulic engineering constructions, and etc. In other words, in these conditions, a restoration of broken natural connections and further deep processing of initial natural systems under the influence of earlier technogenesis products concurrently occurs.

Therefore, two main types of natural system responses—degradation or restoration cause the need for the analysis concerning the development of two directions: different stages and forms of degradation, also the stages and forms of violation restorations of natural systems or their components (Table 2.1).

The increase of TG impacts intensity on natural systems without emergence of essential consequence is possible only to a certain limit. There is a critical area—a

Table 2.1 The main taxon units of technogenic geosystem's geochemical systematization of the Ertis macrosystem

Name of taxon units	Basic principles of identification
Type	Belonging to certain zone signs
Subtype	Biological circulation feature of substances
Class	Features of an air gain and carrying out of pollutants, geochemical specialization of emissions and waste
Sort	Features of water migration
Habit	Geochemical feature of soils and maternal breeds

point of toxicity through which transition leads to the destruction and full reorganization of initial landscape and geochemical systems, and can have catastrophic consequences. In different geographical conditions (even in case of TG influences uniformity) they are reached at different intensity of the operating factors.

During the source removal, influence of TG substance streams or energy per unit area of initial landscape and geochemical systems, as a rule, decreases and, therefore, the metamorphism of natural geochemical processes is weakened, changing the structure of zones influence.

The regularities of combination concerning TG zone influences among themselves, their connections with TG components and initial geosystems are individual for each region and have a certain spatial gradient of changes, so it should be the subject of the study.

Estimates of the geosystems condition, under the influence of techno genesis factors require the development of their ecological-geographical classification. Such classification should be basic in an assessment of condition of natural and anthropogenous environment components and form a basis for acceptance of spatially differentiated nature protection actions.

It is rather obvious, that there are several approaches to the classification creation of the changed technogene geosystems, the main of which are: (1) the geointegrated approach, based on allocation in subgeosystem of spatial complete systems, as results of natural and technogenic factors interaction of landscape shaping, degree of natural processes fracturing; (2) geostructural approach based on the natural and technogenic components combination; (3) ecological or nuclear approach (in A. Yu. Reteyum's point of view), representing in fact zoning of anthropogenous impact in the system types as "technogenic source is the environment".

Since the basic ecological-geographical systematization of technogenic geosystems has not been developed yet, it is expedient to consider the geochemical principles of such systematization, to a certain extent uniting the approaches discussed above and considering environmental pollution as one of the most important sites of technogenic influence. Geochemical approach for technogenic geosystems of higher level is based on the accounting of technogenic loading intensity and natural (natural and technogenic) geochemical situation.

It is expedient for the systematization of technogenic geosystems to fulfill two principal requirements: it should have the close basement with the classification suggested by authors (Kasimov and Perelman 1993) and the existing natural classifications by A.G. Isachenko and V.A. Nikolaev, used further. Especially it is necessary to emphasize the need for the general systematization of the natural and technogenic landscapes, created in a river basin or inflow (subgeosystem) in the conditions of strong technogenic transformation. Systematization of the technogenically changed soils also should be based on the initial principles.

Actually, such approach at the different levels of classification demands the use of various bases and criteria of taxons' differentiation. Therefore, as the bases, anthropogenous (social and production) factors of technogenesis are used on the top of the taxon classification levels of technogenic geosystems, and on the lower there taken natural-conditioned ones (Table 2.1).

The main geochemical feature of industrial, transport and other technogenic influence factors is the formation of technogenic anomalies in various components of the geosystems tied to superficial drain. Contrast and spatial position of anomalies depends on a combination of technogenesis zone's functional structure, the defining nature and level of technogenic influence on environment, and the geosystem-geochemical conditions differentiating this influence. Therefore, geochemical classification of city geosystems should be based on two interconnected factors—technogenic and natural ones.

The leading value is given to the technogenic migration, in many respects determined by the attachment to this or that functional zone, in accordance with feature of which the orders of landscapes are distinguished. Many quantitative parameters of technogenic pollution, and also the transformation and degradation nature of biological circulation are connected with them. Five main types of geosystems are distinguished: (1) park and recreational; (2) agrotechnogenic; (3) intended for building; (4) intended for building and transport; (5) industrial, for which the coefficient of pollutants intake contrast from the atmosphere in comparison with a background, fluctuates from less than 10 in a park and recreational zone, up to more than 30 in an industrial one (Kasimov 1990). These are respectively geosystems of weak, moderate, strong and almost full degradation of biological circulation; however, quantitative criteria for evaluation of degradation are poorly developed.

Within orders, due to water transfer features—carrying out pollutant and geochemical specialization of waste emissions and drains, several groups of technogenic geosystems have been distinguished.

The first three orders represent landscapes mainly of pollutants gain (emission). In their limits geochemical differentiation of geosystems in many respects is defined by local migration of pollutant.

Usually smaller athmotechnogenic strain is experienced by park and recreational geosystems. The role of biogenous migration is still big. Considering influence on population health, especially it is necessary to allocate the geosystems, used for execution of agricultural production (gardens, kitchen gardens), which are under double press of pollutants—athmotechnogenic and agrogene (fertilizers, toxic chemicals).

The other types of technogenic geosystems (Table 2.2) are sources of technogenic emission and the place of partial pollutant accumulation. The type of industrial geosystems depending on production type, the extracted raw materials, a power source and the nature of waste is divided into the geosystems of the certain specialization plants and mines, power plants, tailings dams, dumps, disposal sites (Alekseenko 1990).

For the subtypes division of technogenic geosystems as an integrated criterion serve the levels of separate components pollution and the degree of their danger to live organisms within types.

Water migration of chemical elements for technogenic geosystems is considered at the level of sorts, divided due to the combination of oxidation-reduction, alkaline-acid conditions and types of geochemical barriers in soils profile and

Table 2.2 Types of technogenic geosystems (Kasimov and Perelman 1993)

Levels and danger of pollution*	Types of geosystems				
	Recreational	Agrotechno-genic	Urban	Transport	Industrial and mining
Low ($Z_c < 16$; $P > 200$)	Inundated woods, forest plantings by average pollution	Low pollution (arable land)	Low pollution (small population areas)	Roads of regional value	Small open pits
Average, moderately dangerous (Z_c: soils—16–22, snow—64–128; $P = 250$–450)	Recreational with average pollution	Agrotechno-genic with average pollution (gardens)	Average level of pollution (working settlements)	Average level of pollution	Moderate and dangerous level of pollution
High, dangerous (Z_c: soils—32–128, snow—128–256; $P = 450$–800)	Green space of suburbs	–	Dangerous level of pollution (the small cities)	High level of pollution	Dangerous level of pollution (mine, tailings dam)
Very high, extremely dangerous (Z_c: soils > 128, snow > 256; $P > 800$)	–	–	Very high level of pollution (the large cities)	Roads in the cities	Very high level of pollution (the territory of factories and plants)

Note *Z_c—total indicator of pollution, conventional units (anomaly coefficients sum of separate elements); *P*—size of dust loading, kg/km^2 per day; for convenience, the corresponding communities aren't specified in empty cages

(Methodical recommendations about geochemical assessment of environmental pollution sources 1982; References Methodical recommendations about assessment of the chemical elements pollution in city territories 1982) between the interfaced geosystems with preserving the traditional names in landscape geochemistry.

At the same time, the assessment of transformation of geochemical conditions concerning migration and its forecast under the technogenesis influence has particular importance, that can be considered at the appropriate taxon level (subclass) and reflected in fixing of this or that tendency of geochemical conditions change. It is expedient to consider also oxidation-reduction conditions of ground waters.

In geosystems of technogenesis kernels, the intensive athmotechnogenic intake of substance levels influences a relief on pollutant redistribution. Therefore ideas of autonomy and subordination of technogenic geosystems demand essential modification in comparison with natural analogs. Substantially, the postulate on the negligible size of substance supply from the atmosphere to eluvial landscapes loses meaning, which, obviously, can be used only for background conditions.

Technogenesis factors influence not only water, but also air migration of pollutant, and the accounting provision of their relatively main sources of pollution and the prevailing athmotechnogenic streams are required, along with traditional allocation of mobilization zones, transit and substance accumulation (eluvial, transeluvial, eluvial-accumulative, super aquatic elementary landscapes). As a rule, athmotechnogenic anomalies are connected with windwardly slopes or water separate surfaces, and leeward slopes are less polluted.

On these basis, sorts of technogenic geosystems are allocated (windwardly transeluvial, leeward transeluvial, and etc.). It should be noted, that it isn't clear yet, how better to consider a ratio of natural and technogenic factors in classification. In this systematization, the influence of technogenic factors was considered at higher taxon level (class).

There is an important meaning in implementation of such geosystems into natural (water-erosive) or natural-technogenic (to basins concentration of storm drain) and cascade systems of certain order, and also the openness or isolation of these systems defining features of migration and accumulation concerning technogenesis products.

Many features of water migration, and also pollution levels are closely connected with particle size distribution of soil and ground. So, initial contents of chemical elements and sorption capacity of sands is much less, than at loams, moreover sands are better washed out by an atmospheric precipitation, and etc. The particle size features of soil and ground are considered in technogenic geosystems types division. At the same time, it is important to distinguish natural soil and ground from technogenic soils, deposits and the asphalted surfaces.

The signs and parameters, serving as the separation cornerstone of taxon levels, in case of the appropriate computer processing represent, in fact, the database, necessary for the creation of the ecological-geographical information system concerning the technogenically transformed territories (Geograficheskoe prognozirovanie priroohrannyih problem 1988; Svirezhov 1982; Vladimirov et al. 1986; Volkova and Davydov 1987; Kalygin 2000; Shukputov 2001).

Thus, even in case of quite simple structure of production, the character and the dynamic of TG impacts on the environment are ambiguous, but they can be grouped and ranged on possible response of landscape and geochemical systems or their separate components to them. The signs and parameters, being the separation cornerstone of taxonomical levels, in case of the appropriate computer processing represent, in fact, the database necessary for creation of ecological-geographical information system.

2.2.2   Geochemical Analysis of Technogenic Impact and Factors of Technogenesis

Technogenic pollution began far back in the past. In heritage from the first Altai getters of bronze era, there were career dredging, dumps of overburden breeds, sub-standard ores and factory slags which, being exposed to oxidation processes, continue to pollute the environment. In a "pure" look, ancient technogenic anomalies of bronze era and later are fixed only in those places, where production and ores processing had been thrown in due time and later wasn't renewed. In other fields, which have been involved anew in development during the late period (Zavodskoe, Zmeinogorsky, Zyryanovsky, Orlovskoe, Nikolaevskoe, Riddersky, and etc.), traces of former production have been shaded by more powerful, large-scale developments and have practically not remained. The negative impact of ancient and old enterprises on the environment, their specific weight is rather small.

The main enterprises of mining and metallurgical complex are located in a zone of the most dense river network. Owing to the technical need, the largest enterprises of power system are located here. Such arrangement means, that all pollutants with gaseous, liquid and solid waste from the industrial enterprises inevitably get to river network and soil, causing ecological damage, both to biosensors, and area population.

In regards to the degree of negative impact on the environment, operating enterprises can be arranged in such sequence: (1) metallurgical production; (2) mining and processing factories; (3) pits of open works; (4) underground mines.

Ecological blocks of any industrial city, between which pollutant streams are formed, are conditionally divided into three groups: (a) emission sources, which include industrial city complex, city housing-communal services and transport; (b) transit means, directly accepting emissions, where there is a transportation and partial transformation of pollutants—the city atmosphere, atmospheric losses (rain, snow, dust), temporary and constant waterways, surface water and reservoirs (ponds, lakes, reservoirs), ground waters; pollutants come to these natural systems through the opened and closed collectors by dispersion through the atmosphere or from warehousing of solid waste; (c) depositing environments, in which techno-genic substances—ground deposits, soils (especially, sites of geochemical barriers), plants, microorganisms, city constructions, city population are collected and transformed.

By anomaly degree, concerning lithosphere percentage abundance, the first place is taken by emissions of the enterprises (tungsten, antimony, lead, cadmium, nickel are especially strongly concentrated in the dust), then a little less or comparable to them is loading from waste, drains take the third place in anomaly row. However, due to an absolute lot of delivery to the environment, solid waste advances in emissions. The large number and unevenness of technogenic sources placement in combination with an environment, create a difficult picture of geochemical fields and abnormal zones in the territory of the industrial cities. Identification of technogenic sources in the large city is more complex challenge, in comparison with the separate highly specialized enterprises in small cities and settlements. Therefore, the inventory of technogenic sources is one of the major tasks and priorities in the ecological-geochemical estimation of the cities' condition (Bykov 1988; Tehnogennyie potoki veschestva v landshaftah i sostoyanie ekosistem 1981; Dreyer 1997; Mickiewicz and Sushik 1981).

The major factors, influencing the quality of atmospheric air in large industrial centers of the East Kazakhstan Region (EKR) are emissions of enterprises, and the growing number of pollution from the stationary and mobile sources. The pollutants (P) make negative impact on the environment, at the same time, catching and neutralization of emissions at the level of 90% for the stationary sources have been reached now.

The main part of emissions from the stationary sources fall on the regional center (to 45%), the automobile transport is also observed as there is a significant increase in quantity of mobile sources. The situation is aggravated by the features of atmosphere circulation over Ust-Kamenogorsk, since here for a year they have on average > 100 days with adverse meteoconditions (AMC) during the summer-autumn period, when the calm type of weather prevails.

The greatest strain from emissions of pollutants on the EKR is experienced by the atmosphere of cities such as Ust-Kamenogorsk, Semey, Zyryanovsk and Ridder (Table 2.3 and Figs. 2.2,2.3,2.4), the data characterizing anthropogenous impacts on atmosphere of the EKR cities and regions are submitted.

Quantity of pollution sources in the atmosphere on the EKR makes 17,189, among them organized ones are 10,011, and equipped with treatment facilities are 1797. The high level of pollution is explained by low air out of atmospheric space. The airborne pollutants collect in a ground layer of atmosphere, and their concentration remains at very high level. Load of highways with city transport, complexity of automobile transport exhausts is one of the main sources of air atmospheric pollution by nitrogen dioxide, carbon oxide, organic substances in

Table 2.3 Emissions of pollutants in the atmosphere from the stationary sources of the EKR administrative region settlements in 2012–2014/The program of development of the East Kazakhstan Region territory on (2016–2020)/

Administrative regions	2012	2013	2014
Ust-Kamenogorsk	61.5	55.8	55.7
Semey	27.6	25.6	22.9
Ridder	9.6	6.8	9.1
Kurchatov	1.4	1	1.2
Abay	0.5	0.2	0.2
Ayagoz	2.6	2.5	3.2
Beskaragay	0.5	0.1	0.7
Borodulikhin	3.3	2.5	2.3
Glubokov	3.8	3.6	2.8
Zharmin	3.8	3.7	5.7
Zaysan	1.9	1.8	2.1
Zyryanov	10.7	11.9	12.7
Katon-Karagay	0.3	0.5	0.3
Kokpektyn	0.7	0.7	0.7
Kurchum	0.8	1.1	1.1
Tarbagatay	1	1	0.9
Ulan	1.8	0.8	1.6
Urdzhar	2.6	1.8	1.9
Shemonaikhy	5	2.6	3.5

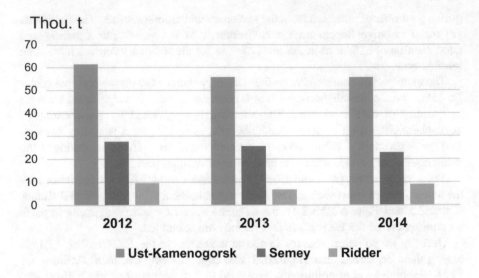

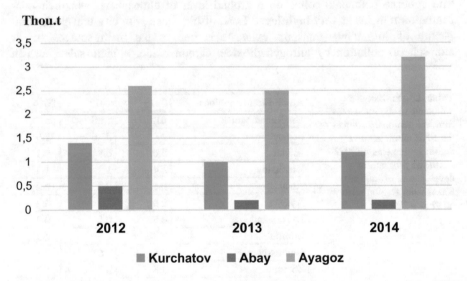

Fig. 2.2 Emissions of pollutants in the atmosphere from the stationary sources of the EKR administrative region settlements in 2012–2014 (The program of development of the East Kazakhstan Region territory on 2016–2020)

settlements, and moreover high load of highways even with a good air out, leads to the accumulation of harmful impurity in the air atmosphere.

High level of air pollution in the EKR is caused by the emissions of enterprises of nonferrous metallurgy, power system and automobile transport, and also by the climatic conditions, not favorable for pollutants dispersion. Pollutants composition in emissions on the EKR contains up to 170 names, from them 22% belong to 1

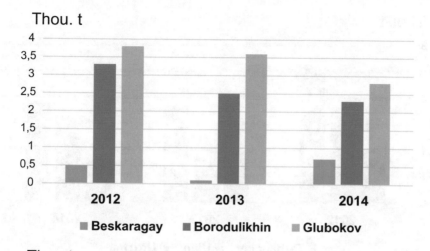

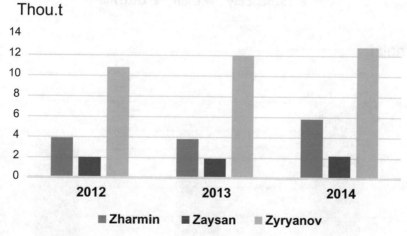

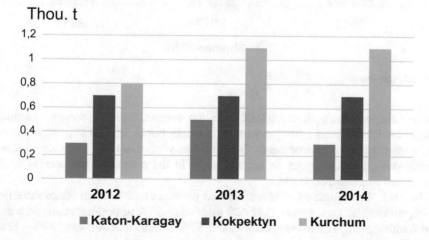

Fig. 2.2 (continued)

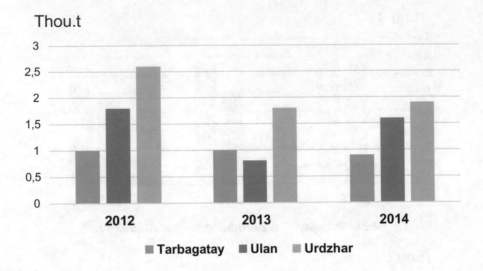

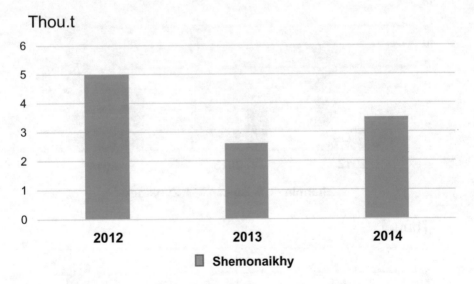

Fig. 2.2 (continued)

class of danger (among them are lead, cadmium, arsenic, fluoric hydrogen, chlorine, beryllium, which gross emission percent is small, but their high toxicity is extremely dangerous to environment). Besides, many of them have the summation effect, strengthening impact on human health in the combined presence of such substances in atmospheric air.

In 2013, the volume of resolved limit on pollutant emissions in atmosphere for enterprises of the 1st category (147 630 tons/year): 33% of pollutant emissions are the metallurgical branch enterprises share; 13%—municipal services; 17%—heat

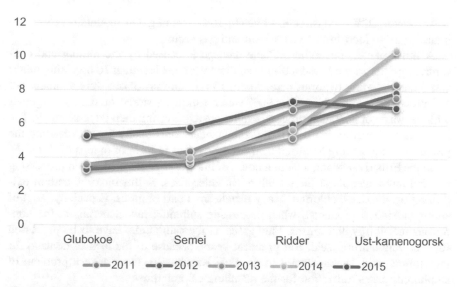

Fig. 2.3 Dynamics of the atmosphere pollution index of the EKR (APS5) for individual cities/Environmental protection and sustainable development of Kazakhstan 2011–2015/Statistical collection/Committee on statistics of the Ministry of national economy of the Republic of Kazakhstan/

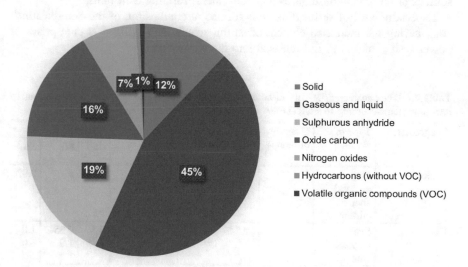

Fig. 2.4 Emissions of the widespread substances of the EKR departing from stationary sources in 2015 which are most polluting the atmosphere/Environmental protection and sustainable development of Kazakhstan 2011–2015/Statistical collection/Committee on statistics of the Ministry of national economy of the Republic of Kazakhstan/

power sector; 22%—construction branch; 10%—mining and mountain processing industry; 2%—food industry; 1%—oil and gas sector.

A little smaller, but rather notable damage is caused by the mining and concentrating enterprises. In the recent past, there were not less than 16 operating mines and 9 concentrating factories here. About 13 million tons of ore were annually got and processed, from which over 90% weren't utilized, stored in storages together with the waste of floatation process, even more aggravating their adverse effect. The huge mass of overburden breeds and sub-standard ores, many times exceeding the produced ore are being localized near the enterprises in form of dumps (Table 2.4).

In the Ertis river basin, a large amount of flue gases is released, when processing mineral raw materials at the metallurgical enterprises. Sulfur dioxide, carbon oxides, nitrogen oxides, chlorine, heavy metals are a part of gases. Especially, a lot of sulfur dioxide is produced, when processing sulphidic raw materials at the non-ferrous metallurgy enterprises. Flue gases, at the same time, have to be processed for receiving sulfuric acid. But in recent years, because of the lack of demand for the sulfuric acid, manufactured in the East Kazakhstan, there arise a problem of sulphurous gases utilization for the metallurgical enterprises.

Metallurgical enterprises are also the main supplier of heavy metals emission. In turns of zinc-lead production, a large number of arsenic combinations released into the atmosphere, which have to be buried, circulate. The enterprises of power system and automobile transport as well as the metallurgical enterprises are found to be the sources of the greenhouse gases and substances, forming acid rains.

Especially a lot of sulfur dioxide is formed at combustion of the Semipalatinsk coal, having the increased content of sulfur, which isn't utilized at enterprises of power system and is in full released into the atmosphere.

Table 2.4 The characteristic of metallurgical production waste in the Ertis-Shulbinsky intrageosystem/according to EKRTDEP/

Subgeosystem	Type of waste	Quantity of dumps	Total amount, one thousand m^3	Area of alienation, sq.km	Contents, %		
					Cu	Pb	Zn
Shulbin	Slag disposal area, clincer	1	1500	18	0.45	0.75	0.92
Uba	Slag disposal area	5	12,340	Data is absent	0.5–2.45	0.35–1.8	1.0–6.0
	Dumps of solid wastes	1	637	Data is absent	Data is absent		
Ulba	Slag collection system	3	923	The same	The same		
Total		9	15,400				

Problem of pollutant emissions in the atmosphere isn't located in the place of their emergence, as the cross-border transfer of atmospheric industrial emissions of high concentration is possible at the distance ranging from 400 to 500 km, where they drop out in the form of acid rains, are besieged on the land surface, get to water, acidify the soil, harm elements of biocenoses.

As for the atmosphere pollution of cities, major striking factors are connected with the enterprise's activity of metallurgical complex and power system. There are steady excess emissions of sulfur dioxide, nitrogen oxides, phenol, bioxyde format, and lead. Volumes of dust emissions, carbon dioxide, chlorine, arsenic begin to increase. As a result, there is a direct impact on the biosphere in form of emissions of physically and chemically active agents, inert materials, and there come a direct heating of the atmosphere and physical impact in the form of acid rains.

The analysis of emissions placement on given areas shows that they are divided into two groups. The first one includes cities such as Ust-Kamenogorsk, Semey, Ridder, Zyryanovsk, and also the regions of the East Kazakhstan, having the large industrial enterprises, generally mining and metallurgical. Emissions nature for 70% is presented by gaseous and liquid form of pollutants. Main emission of solid pollutants goes from unorganized sources, such as dumps, tailing—and output ponds, raw material warehouses. Main suppliers of pollutants in a gas and liquid form are the metallurgical and power system enterprises.

Other regions of the East Kazakhstan and Kurchatov city make the second group. Emissions nature in the form of the thrown-out substances phase is halved on solid and gaseous the together with the liquid phase. Efficiency of cleaning makes no more than 21%. Main emission suppliers are small boiler rooms without systems of dust and gas catching. The greatest emissions number from automobile transport comprise the share of Ust-Kamenogorsk and Semey (34 and 23.5% respectively).

One of the most effective remedies in dust suppression is gardening. At combustion of one liter of fuel in the automobile engine about 200–400 mg of lead get to air. During the vegetative period one tree can save up such amount of lead, which is contained in 130 L of gasoline. Simple calculation shows that neutralization of harmful operation of one car requires not less than 10 trees. The number of vehicles in Ust-Kamenogorsk annually increases by 5–7 thousand units. For 01.01.2015, the number of vehicles made 105,521 units. Automobile transport becomes the main source of atmospheric air pollution. In the sum of atmospheric air pollution the share of transport emissions makes about 40–60%. The ecological safety center of Ust-Kamenogorsk since 2012 has been exercising visual control of "the smoking transport" with the subsequent data transmission to public authorities for taking administrative measures. The use of low-quality hydrocarbon fuel and vehicles operation for more than 10 years is considered to be one of the main constant pollution reasons.

Measurements of atmospheric air pollution levels in the East Kazakhstan are regularly carried out in Ust-Kamenogorsk, Ridder and Zyryanovsk. Dust, sulfur dioxide, carbon, nitrogen, chlorine, formaldehyde, phenol, arsenic, leads are found. The most toxic of listed ingredients is lead (the 1st class of danger). Dynamics of average annual impurity of atmospheric air in the listed settlements on the atmosphere pollution index (API) fluctuates from 11 to 24 (Cherednichenko and Nedovesov 1997).

In ecologists opinion, it was unfair, that only for the first half of 2016 according to PLT "Kazgidromet" the number of excesses over 1 MPC on this ingredient has made 74 times. For comparison: for the entire period of 2015 the excess of the weighed substances was observed only 34 times.

In the spatial distribution of aero technogenic streams, the major role is played by geomorphologic factors: polluted air masses are localized either in closed hollows (Zyryanovsk), or move due to the daily inversion lengthways the steep-sided valleys of large water currents (Ertis, Ulba, Bukhtyrma). The water separate ridges framing them are natural barriers, interfering with wider vulgar circulation of aero technogenic loops. The most powerful aerogenic stream functions along the Ertis river valley. In the northwest direction its influence is felt to border with the region Semey, in the southeast—to the Bolshenarymsky village; not less stable aero technogenic and water streams are observed along the Ulby and Bukthyrma valleys.

The carrying out of dust into the atmosphere from dumps of the Ridder Polymetallic Plant (RPP) fluctuates from 0.5 to 1.5 thousand hectares that averages 100 tons a year. Integrated calculations show that dusting from destruction of all technogenic waste in a year makes 113 thousand tons, and for the entire period of mining production dust demolition only in Ridder district makes 300 tons in a year.

According to "The National center of health and work-related diseases", which conducted meteorological researches in the urbanized territories in 2015, in Ust-Kamenogorsk the excess of dust in average daily MPC has made 64%, lead, chrome, copper, zinc, cobalt, arsenic being its chemical compositions.

The highest dust loadings are indicated in the zones of the large industrial enterprises of mineral and raw complex. For example, in the Gluboky settlement (Ertis copper smelting plant ECSP) they reach 2560 kg/km^2 per day, in Ust-Kamenogorsk (Lead-zinc plant LZP) is 2987 kg/km^2, in Ridder (RPP)— 1000 kg/km^2 per day, and more, that corresponds to extremely dangerous extent of atmosphere pollution (Fig. 2.3).

In areas of mining and processing productions (intended for building zones of Zyryanovsk, the Belousovka settlement, and Verkhneberezovsky) dust loadings has reached 700 kg/km^2/day. Metal loadings (in estimates on Zindicator) in the specified epicenters of pollution make 400–700 excesses over a single background that corresponds to a very strong level of atmosphere pollution. Atmospheric air conditions in the region, especially in settlements—the centers of industrial production, it is necessary to recognize as crisis.

Atmospheric air in the cities is usually polluted by sulfur oxides, nitrogen, and dust, but the increased concentration of pollutants are especially dangerous, being specific to each type of production. The highest levels of pollution are observed in the cities with black, color and petrochemical industry where MPC of harmful substances are already exceeded several times. Among specific pollutant in the cities, prior positions are taken by the polycyclic aromatic hydrocarbons (PAH), formaldehyde and heavy metals. Especially contrasting are technogenic anomalies

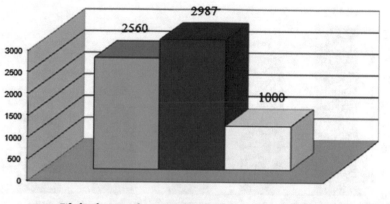

▫ Gluboky settlement (Ertis copper smelting plant ECSP)
▪ Ust-Kamenogorsk (Lead-zinc plant LZP)
▫ Ridder city (Ridder Polymetallic Plant RPP)

Fig. 2.5 Dust loadings from the industrial enterprises of kg/km

of one PAH—3,4 benzpyrenes, having cancerogenic properties and formed, mainly, when burning the fossil fuel (Fig. 2.5)/according to EKRTDEP/.

In the cities dust content of air is high. So, in background geosystems the supply of solid substance from the atmosphere makes 10–15 kg/km per day.

In the industrial cities, it increases by 5–10 times and more, that leads to growing up role of the weighed particles as carriers of chemical elements and contrast of the technogenic anomalies, formed in atmospheric drop-outs. At the same time two types of athmotecnogenic loading are distinguished: (1) loss of dust in large amounts with rather low concentration of pollutant and (2) the high loadings formed by loss of smaller amount of dust with the increased chemical element contents.

Because of many pollutants global distribution, especially heavy metals, there are great difficulties when determining a regional background of atmospheric drop-outs. Soils are the main supplier of naturally originated heavy metals in the atmosphere. It is considered that for these metals, and also arsenic and antimony, the anthro-pogenous contribution to their total amount in atmosphere already makes more than 50%. Therefore, the concept "background" for atmospheric losses is relative. On a regional background of loss in the industrial cities, on average at 3–15 times are enriched with heavy metals. In turn, the territory of Ust-Kamenogorsk is, as a rule, polluted unevenly and on the raised city background stands out clearly that the technogenic anomalies of losses dated for industrial zones, where concentration of zinc, lead, nickel, mercury, chrome and other metals increase usually by 5–6 times.

Intensity of air pollution in the cities depends on the number of physiographic factors and, first of all, on meteorological situation and land relief. Especially strong pollution is characteristic for the industrial cities, located in the mountain hollows (mountain-kettle community) with frequent inversions of temperatures.

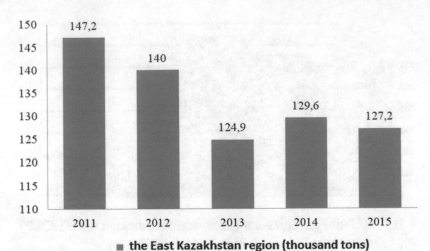

Fig. 2.6 Volume of emissions in the EKR

The sources of technogenic and anthropogenous impact on the environment of Ertis-Shchulbinsky subgeosystem in Ust-Kamenogorsk are the enterprises of non-ferrous metallurgy, power system, food and processing industry, municipal enterprises, automobile and railway transport.

The most ecologically problem enterprises are: JSC "Kazzinc", JSC UMP, AES Ust-Kamenogorsk HPP, AES Sogrinsky HPP, Ust-Kamenogorsk TMP.

According to the Statistics department of East Kazakhstan region for 2015 year, 18 592 sources of atmosphere pollution are registered, from them organized ones are 10 306. In the city Ust-Kamenogorsk, 5 899 sources are registered, out of them organized ones makes 3 324. The volume of emissions from stationary sources to the atmosphere in 2011 made 147.2 thousand tons per year. From 2012 to 2015 year decrease is observed: in 2012—140.0 thousand tons, in 2013—124.9 thousand tons, in 2014—129.6 thousand tons, in 2015—127.2 thousand tons (Fig. 2.6)/according to the Ministry of national economy of the Republic of Kazakhstan, Committee on statistics/.

The main sources of pollutants with nitrogen dioxide, sulphurous anhydride, formaldehyde, benzapyrene, phenol, carbon oxide and the weighed substances are the metallurgical and thermal industry enterprises, such as UK MC JSC "Kazzinc", JSC "AES Ust-Kamenogorsk HPP", JSC "AES Sogrinsky HPP" JSC "Ust-Kamenogorsk Thermal Networks".

About 80% of all emissions in the atmosphere in the area are the share of the cities Ust-Kamenogorsk, Semey, Zyryanovsk, and Ridder. Dynamics of atmosphere pollution index (APS_5) around the city Ust-Kamenogorsk from 2011 to 2014 is the following (Table 2.5).

For 2014 in general around the city, the level of atmosphere pollution belongs to high pollution, the IIIrd gradation. It was defined by the LF (largest frequency) value of equal 28% (high pollution), the SI (standard index) = 9 (high pollution).

Table 2.5 Dynamics of atmosphere pollution index (APS$_5$) around the city Ust-Kamenogorsk/according to the Ministry of national economy of the Republic of Kazakhstan, Committee on statistics/

Year	2011	2012	2013	2014
APS$_5$	8.4	7.9	7.6	9.5

During the period from 2011 to 2014, the general gross emission of atmosphere pollutants decreased from 147.2 thousand tons to 129.6 thousand tons, around the city Ust-Kamenogorsk it has decreased from 61.5 thousand tons to 55.7 thousand tons.

The total *amount of industrial emissions in atmospheric* air from the large enterprises in 2015 made 101,91,361,585 thousand tons, that is for 1,740,076,515 thousand tons (for 2,02%) less in comparison with the volume of emissions in 2014.

- volume of sulphurous anhydride emissions—56,723435 thousand tons;
- volume of nitrogen dioxide emissions—21,123432 thousand tons;
- volume of solid emissions—13,1,357,098,962 thousand tons;
- volume of carbon monoxide emissions—10,931038954 thousand tons (Fig. 2.7).

The reduction of emission volumes is caused by the fact that limits have decreased in comparison with 2014, and also there was reduction in the emission volumes of the large enterprises such as:

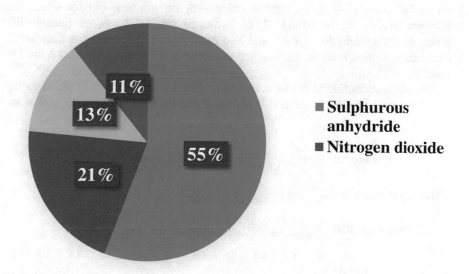

Fig. 2.7 The volume of industrial emission into atmospheric air from large enterprises for 2015

– LLP "Kazzinc" ZMPC (Zyryanovsky Mining and Processing Complex)—reduction of pollutant emissions in atmospheric air is connected with reduction of working hours of limy plant furnaces;
– LLP "Kazzinc" UK MC—reduction of pollutant emissions by 15% in comparison with last year is proved by stage-by-stage development of new Isesmelt technology at the reconstructed lead plant instead of agglomeration;
– JSC "Vostokmashplant"—decrease in production output;
– LLP "Ulba fluorine complex"—reduction of processed ore volumes, lack of overburden and mining works;
– LLP "Artel Diligent Miner"—decrease is caused by suspension of mining operations;
– LLP "Silicate"—small volume of products realization;
– LLP "Semey alloy"—lack of raw materials.

Number of the enterprises such as: LLP "Bari-B and K", LLP "Avtodorservice firm", JSC "Bast", LLP "Samar Astyk", LLP "Kvarta", LLP "Murager PCF", LLP "Pribrezhny-1", MUS "District Heating Company" don't carry out production activity.

For 2015, according to the **stationary** network of RSE "Kazgidromet" observations, the city Ust-Kamenogorsk is characterized by the high level of pollution. In general, around the city, average concentration have made: sulfur dioxide—1, 6 shift-average MPC, nitrogen dioxide—1, 2 shift-average MPC, ozone—1, 8 shift-average MPC, other pollutants—didn't exceed MPC (Fig. 2.8). The number of excess cases more than 1 maximum one-time MPC on the weighed substance is 37, on sulfur dioxide—790, on carbon oxide—78, on nitrogen dioxide—312, on nitrogen oxide—12, on ozone—4, on hydrogen sulfide—5849, on phenol—80 cases, on formaldehyde—1 case, and also more than 5 maximum one-time MPC excesses were observed on the weighed substance and hydrogen sulfide on the 1st time.

For 2015, according to the fixed stationary network observations of atmospheric air of the city Ridder, in general is characterized by the high level of pollution.

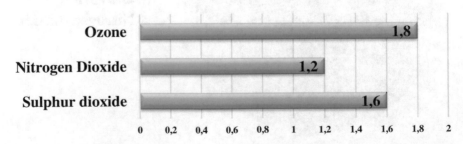

Fig. 2.8 The average concentration of pollutants in the Ust-Kamenogorsk city for 2015 (MPC)

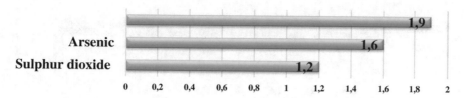

Fig. 2.9 The average concentration of pollutants in the Ridder city for 2015 (MPC)

Altogether, around the city average concentration has made: ozone—1, 9 shift-average MPC, arsenic—1, 6 shift-average MPC, sulfur dioxide—1, 2 shift-average MPC, other pollutants—didn't exceed MPC (Fig. 2.9). Excess cases more than 1 maximum one-time MPC have been registered on sulfur dioxide—64, on nitrogen dioxide—32, on nitrogen oxide—72, on hydrogen sulfide—6894, on ammonia—47, on phenol 1 case, it was also observed more than 5 maximum one-time MPC on nitrogen oxide—32, on ammonia 13 times.

For 2015, according to the fixed network of atmospheric air observations of the city Semey is characterized by the increased pollution level. In general, around the city average concentration has made: ozone—1, 3 shift-average MPC, phenol—1, 9 shift-average MPC, other pollutants—didn't exceed MPC (Fig. 2.10). Number of excess cases more than 1 maximum one-time MPC have been revealed on the weighed substances-1, the weighed particles RM-2.5–275, the weighed particles RM-10-173, on carbon oxide—14, on nitrogen dioxide—339, on nitrogen oxide— 10, on ozone—32 and on hydrogen sulfide has made 2238, on ammonia—12 cases, also more than 5 maximum one-time MPC on the weighed particles RM-2.5–3 and on the weighed particles RM-10-2 cases.

For 2015, according to the fixed network of atmospheric air observations Glubokoe settlement, in general, is characterized by the increased pollution level. Altogether, on the settlement there was an average concentration of ozone 4, 4 shift-average MPC, other pollutants—didn't exceed MPC (Fig. 2.11). Excesses more than 1 maximum one-time MPC were observed to the weighed particles

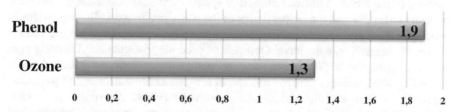

Fig. 2.10 The average concentration of pollutants in the Ridder city for 2015 (MPC)

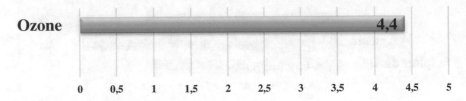

Fig. 2.11 The average concentration of pollutants in the Ridder city for 2015 (MPC)

RM-2.5–71, to the weighed particles RM-10-32, on carbon oxide-1, on nitrogen dioxide—97, on ozone—4927, on hydrogen sulfide—283, on phenol—14 and on ammonia—4 cases.

The region has the high potential of natural resources, promoting broad development of industrial sector of economy. As a result, it bears for itself the whole range of environmental problems.

Following the results of 2014, at the republican level due to the emissions in the atmosphere pollutants, released from the stationary sources of the East Kazakhstan, the region is in the third place, by the quantity of pollutant emission sources—in the second. The volume of formed pollutants has exceeded 1.7 million tons per year.

At the same time, emissions of the substances polluting atmosphere, and releasing from stationary sources (per capita, kg) is equal to the 92.8 or 9th place in the RK (in the first place is Pavlodar region—807.4 kg).

Moreover, the tendency to decrease (Table 2.6) was outlined in the total amount of pollutant gross emissions.

Industrial and municipal activities lead to the considerable technogenic transformation of water balance in their territory. Along with the changes of hydrogeological conditions (flooding, drainage, sag and so forth) one of the main forms of the urban environment's technogenic deformation is the pollution of surface and underground water industrial and household drains. Therefore, hydro geochemical researches represent the necessary block of the urban environment complex analysis. In the territory of the city it is possible to differentiate the following main directions in an assessment of water streams pollution, characterizing water recirculation of the city as a difficult migratory system. The first—defining the structure of sewer industrial and municipal drains as the waste integrated indicators, which is coming in the liquid state to the surrounding urban environment and having various degree of cleaning completeness. Quite often, even the so-called conditionally pure drains contain the high concentration of pollutants, in ten and hundred times exceeding maximum permissible and are, in turn, an additional source of pollution, especially if they are dumped in open reservoirs (lakes, the rivers, reservoirs) coming to sewer network, collector channels, settlers, and etc. The chemical composition of such drains reflects an overall picture of the urban area condition.

Table 2.6 The emissions of the atmospheric air pollutants in the section of the EKR cities and regions (2012–2014 year)/according to the Newsletter in the environmental state in the Republic of Kazakhstan for 2014. Astana: RSE "Kazgidromet", 2014. 213 P/

	Quantity of stationary sources of pollutant emissions, unit	Emissions of atmospheric air pollutants, thousand tons				Region share in the total amount of emissions for 2014, %
		2012 г.	2013 г.	2014 г.	Changes in 2012–2014 years	
the EKR	18,592	140.1	124.9	129.6	−10.5	100
Ust-Kamenogorsk	5899	61.5	55.8	55.7	−5.8	43.0
Semey	4240	27.6	25.6	22.9	−4.7	17.7
Ridder	811	9.6	6.8	9.1	−0.5	7.0
Kurchatov	199	1.4	1.0	1.2	−0.2	0.9
Abay	35	0.5	0.2	0.2	−0.3	0.2
Ayagoz	514	2.6	2.5	3.2	0.6	2.5
Beskaragay	239	0.5	0.1	0.7	0.2	0.5
Borodulikhin	704	3.3	2.5	2.3	−1.0	1.8
Glubokov	706	3.8	3.6	2.8	−1.0	2.2
Zharmin	832	3.8	3.7	5.7	1.9	4.4
Zaysans	502	1.9	1.8	2.1	0.2	1.6
Zyryanov	1366	10.7	11.9	12.7	2.0	9.8
Katon-Karagay	63	0.3	0.5	0.3	0.0	0.2
Kokpektyn	201	0.7	0.7	0.7	0.0	0.5
Kurchum	203	0.8	1.1	1.1	0.3	0.8
Tarbagatay	95	1.0	1.0	0.9	−0.1	0.7
Ulan	319	1.8	0.8	1.6	−0.2	1.2
Urdzhar	389	2.6	1.8	1.9	−0.7	1.5
Shemonaikhin	1275	5.0	2.6	3.5	−1.5	2.7

At an adverse condition of the sewerage, they can also be a secondary source of pollution, mainly, of underground waters. The hydrogeochemical features of superficial drain waters in the cities significantly differ from background conditions. The chemical composition of waters, the degree of their mineralization, the contents and the ratio of macro components are changing. Researches of E.P. Yanina in Moscow, Podolsk and other cities, also researches of N.A. Barymova in Kursk, have shown that low-mineralized (200–400 mg/l) hydrocarbonate background waters in the cities become saltish (1 g/l and above), hydrocarbonate-sulfate, and during snowmelt, when deicing mixes are dissolved,—chloride sodium. On average in city superficial drains the content of chlorine, sulfate–nitrite—and fosfation, sodium and potassium in ten and hundred times more, than in background conditions. At the same time, on the firm asphalted surface the ionic drain sharply increases, which is 10 times more, than in natural and agricultural landscapes.

The presence of synthetic pollutants—phenols, oil products, the superficially active agents (SAA), polychloride biphenyls (PCB),—in some cases the heavy metals strengthening migration, due to the formation with them the soluble complex connections is especially characteristic for these waters. Therefore, unlike background waters in the polluted surface water of the city, there is an increase of soluble, mainly the organic forms of cadmium and nickel forming with SAA rather steady helatny combinations.

On the contrary, for mercury, copper, zinc and lead the share of technogenic suspension increases, in which they are mainly in geochemical mobile sorption and carbonate, organic and hydrooxidic forms that allows them to join the technogenic migration.

In this regard, particular importance is given to the third direction of hydrochemical estimates of the urban environment—studying of the final links condition of water recirculation waste and surface storm water (river and underground waters), the quality of which is deteriorating as a result of technogenesis, as well as the ground deposits (technogenic muddy material), serving as the integrated indicator of reservoirs technogenic load in city boundaries.

The most part of the Big Altai territory belongs to the Ertis river basin. The main contribution to the pollution of this water system is made by metallurgical and mining and processing complexes, and also by the municipal services of the cities and working settlements (Table 2.7).

In the East Kazakhstan for 2014, there were 174 enterprises of water users; from them 48 enterprises have dumpings into superficial reservoirs. Water resources of the Ertis river basin are being exposed to the greatest pollution. The main sources of

Table 2.7 Pollution by reservoir metals in a zone of influence of the area's mining enterprises/according to EKRTDEP/

Water objects	Excess of toxic metals MPC mg/l^3			Pollution sources
	Copper	Zinc	Oil products	
The Breksa River	20–24	25–30		Mine and drainage waters of the Shubinsky minery, Talovsky tailing dams and minery of SMR Ridder MPP JSC "Kazzinc"
The Ulba River	10–12.5	10–12	2.5–2.8	Drainage waters of dump №2 of the Tishinsky minery, MPP Ridder
The Glubochanka River	10–13	20–22	2–3	Belousovsky minery and concentrating factory of "Vostokkazmed" branch of Kazakhmys corporation
The Uba River	2–3	3–4	3–4	Influence of the unfinished minery Chekmar and Nikolaev pits, Snegirikhinsky minery

pollution are naked surfaces of excavations, dumps, tailing dams and product dams of concentrating factories, dump products and industrial drains of metallurgical, chemical-metallurgical, chemical, heat power and machine-building enterprises. In coincidence with the existing assessment, the ecological damage from pollution of the ground waters by the mining and concentrating enterprises and industrial dumps in Ust-Kamenogorsk makes 86 million US dollars.

The rivers Ertis and Uba within the studied region, have the 3rd class of water quality—moderately polluted with the main contribution from not enough purified mine waters, dump drainage waters and industrial waste waters. Pollution by copper, zinc, nitrogen nitride is being observed in the river Ertis. The Ulba is polluted by copper, and zinc. Because of the critical situation with the purification of city sewage, connected with a lack of treatment facilities design, capacity and prolonged terms of construction concerning the third turn treatment facilities (it has begun since 1989), insufficiently purified sewage to 200 thousand m^3 per day is dumped. The extensive area of underground water pollutions was created at the interfaced industrial sites of UK MC, UMP, UK CHPPs and adjacent to them from the north site is tailing dams of UMP. The level of underground water pollution in the area borders has reached on zinc, copper, lead, cadmium, selenium, manganese, arsenic of tens and hundreds of MPC, on fluorine, beryllium of 200–300 MPC, nitrates and nitrites of 3–30 MPC, nitrogen of ammonium 50 MPC (the UMP profile, an industrial site and tailing dams).

The river Bukhtyrma, having the 2nd class of water quality is pure, however after confluence of the Beryozovka river, the class of quality has decreased to the 3rd one. It is connected with the dumping of not enough purified mine waters of the Grekhovsky minery and concentrating factory of the Zyryanovsky MPP.

The rivers Breksa, Tikhaya, and Ulba, have respectively the 6th and 7th classes of water quality—extremely dirty and very dirty. In the river of Breksa, there is copper and phenol. In the river Tikhaya, there is copper, zinc, phenol. In the river Ulba, there is also copper, zinc and phenol. The main impact is exerted by mine and drainage waters of the Shubinsky minery, Talovskoye tailing dams and the SMR mineries, drainage water dumps of the № 2 Tishinsky minery of Ridder MPP JSC "Kazzinc".

The Krasnoyarsk rivers, which have the 7th class of water quality, are extremely dirty. The significant contribution is made by crude mine waters of Verh-Berezovsky and Ertis mineries. There is high excess of MPC on zinc, copper, cadmium, and manganese. The Glubochanka river, having the 6th class of water quality, is very dirty. Main contribution: Belousovsky minery, EMP, refuse-fired plant of the Glubochanka settlement. It has local excess of MPC on copper and nitrogen nitride. It is lower than merge of the Ertis river and the Krasnoyarsk river, there is excess of MPC on copper, zinc, nitrogen nitride.

In the dynamics of dumping change concerning the qualitative composition of the basic pollutants into the EK reservoirs on copper, lead and cadmium, the main contribution was made by the JSC "Kazzinc" (67,9, 98, 64.4% respectively), on zinc—the Belousovsky MPP (54.8%) and the Berezovsky minery (34.6%).

Pollution by oil products is in many respects connected with a spill of aviation fuel in the Semey suburbs.

The Belousovsky concentrating factory in 1990 was discharged from the construction of the new tailings dam, that has considerably complicated warehousing of the current tails, which aren't processed in any way, and dry warehousing of tails for the filling purpose of the developed space in mining has become complicated because of maintenance excess, involving soluble forms of non-ferrous metal minerals and toxic combinations.

The considerable pollution by mineral raw materials of geosystems is happening to the drainage and insufficiently purified and not purified mine waters. Because of the economic crisis, a part of fields were thrown, without accomplishing operation events, currently becoming the essential pollution source. This is Chekmar, Berezovsky, Novo-Berezovsky, Yubileino-Snegirikhinsky, Pokrovsky, Rulikh, Sugatovsky, Belogorsk, Bakenny, Maralikhinsky, and Shemonaikhy.

Due to the forthcoming working off of the stocks, the preservation of the following fields as Zyryanovsky, Grekhovsky, Akzhal, Dzherek, Mukur, Tishinsky, Bakenny, Belousovsky, Nikolaev, Ridder-Sokolnoye, Bakyrchik, Shubinsky, Zheskent, and Bolshevik will be required. Pollutants get to river system and soil, from there they go to the elements of biocenoses.

Problems with the household drains are connected generally with the shortage of power treatment facilities in Ust-Kamenogorsk and Semey, and also with the "Vodokanal" enterprise in Beryozovka settlement and condition of treatment facilities of the household sewerage in Nikitinka settlement, "Riddersky"sanatorium, Sekisovka village, Opytnoe pole settlement, Saratovka settlement, and Kozhokhovo settlement.

Besides, cadmium, copper, lead, zinc, selenium, thallium, and also arsenic are recorded in sewage. Their total annual volume in dumpings fluctuates from 26 to 57 tons/year. Further, downstream the Ertis accepts dumpings of Zyryanovsky lead plant (the Bukhtyrma river), the Belogorsk OMPE, CRCE (Casting and rolling complex enterprise), SCC (Service of central communications), Pipe Metallurgic Company (the Tikhaya and Ulba rivers), YCSP (the small rivers Glubochanka and Krasnoyarsk), industrial and municipal drains of Semey, which are "transported" then to Russia.

The Ertis river within the Rudny Altai is exposed to the most intensive pollution by heavy metals, where large technogenic biogeochemical province over 12.5 thousand m^2 was created. Water-soluble or firm products of drain are formed in the Ertis basin, which are deposited, either in final accumulation zone, or in transit zone. It should be noted, that there is a close hydraulic connection of river waters with the developed water-bearing horizons in river valleys. The rivers and specified water-bearing horizons of river valleys practically represent united water system.

Under natural conditions, the rivers can be drains and at operation of borehole riverside water intakes (infiltration) type, feed the water-bearing horizons and participate in the formation of the underground water's operational reserves. For the specified reason, protection against bacterial and chemical pollution of surface water can provide not only a normal sanitary and hygienic state of water currents and reservoirs, but also the required quality of water on water intakes from wells.

The technogenic influence danger to the person is connected with fact that the river Ertis is a source of water supply of 550 large settlements, more than 47 industrial and agricultural enterprises, and is used for irrigation of more than 247.4 thousand hectares of lands and 407.1 thousand hectares of jellied hay makings and estuaries.

Because of the enormous water area, the Ertis basin under technogenic influence has the majority of the Southwest Altai geosystems. Water pollution of the Ertis has a cross-border character; so, in an alignment near the Buran settlement, increased copper concentration (4 MPC) and oil products (5.5 MPC) are already noted, and it is caused by river pollution in the China territory as a result of copper-nickel plant activity.

In surface water of next areas of warehousing of RPP waste it is found in mg/dm^3: Pb 0.24–0.7; Zn 110–200; Cu 1.5–150.

One of the main surface water pollutants of the Ulba subgeosystem is the waste water dump of № 2 Tishinsky minery. Average annual debit of this water is equal to 150 m^3/h, yearly average content of zinc, copper and cadmium respectively mg/dm^3: 158, 1.5, 0.5. It considerably exceeds the admissible concentration of metals in water, especially on zinc (to 50 threshold limit value MPC).

Drain of the river Uby isn't able to self-clean, and pollution reaches its mouth, near Ust-Kamenogorsk, where the concentration of Zn makes 0.019 mg/dm^3 (1.9 threshold limit value MPC), Cu 0.008 mg/dm^3 (8 threshold limit value MPC).

Maximum concentration of heavy metals (copper, zinc) in open waterways is noted in the places of monitoring, located close to the operating enterprises of mineral and raw complex: the river Ertis—1 km is lower than the river confluence Krasnoyarka, the river Bukhtyrma—5.9 km is lower than mouth of the river Beryozovka, the river Ulba—near the Tishinsky minery (Table 2.8).

Despite some decrease in concentration of harmful substances in the river waters, which is connected with the production stagnation, their impurity **over time** remains rather high. Dumpings into the small rivers as Breksa, Ulba, Krasnoyarka, Glubochanka, and etc. are especially pernicious.

Table 2.8 The river pollution indexes/according to the Newsletter on state of environment in the Republic of Kazakhstan for 2014. Astana. RSE "Kazgidromet", 2014. 213 P/

Name of water object, point	2014	2015
the river Ertis (East Kazakhstan)	1.13	1.86
the river Ertis (Pavlodar)	0.96	1.8
the river Bukhtyrma (East Kazakhstan)	1.17	1.46
the river Ulba (East Kazakhstan)	5.20	5.0
the river Breksa (East Kazakhstan)	4.65	8.49
the river Tikhaya (East Kazakhstan)	7.88	6.08
the river Glubochanka (East Kazakhstan)	4.59	7.54
the river Krasnoyarka (East Kazakhstan)	8.54	11.51

The main reason for intensive pollution by sewage lies in the absence or imperfection of the existing treatment facilities, insufficient introduction of technologies with the closed water recirculation.

So, on the Rudnoaltaysky region, the amount of water in reverse and reuse makes only 34–38%, on the region Semey—3.3–6.8, and 80–90% considered as necessary.

The technogenic pollution of underground waters is rather localized and formed on sites of the enterprises activity, connected with the mineral and raw complex, power system, and agriculture. Underground waters of the considerable part of the Rudnoaltaysky region are polluted by toxic components. Especially, critical situation has developed in settlements, which are the centers of industrial production that makes significant problems of drinking water supply for population.

In comparison with 2014, the quality of water in the rivers Kara Ertis, Ertis, Bukhtyrma, Tikhaya, Ulba, Oba, Emel, Ayagoz, lake Markakol, Bukhtyrma reservoir, Ust-Kamenogorsk hasn't changed significantly, but in the river Breksa, Glubochanka, and Krasnoyarka it has worsened.

In territory of the EKR, extremely high and high pollution are recorded in the following water objects: the river Ulba—27 cases of HP, the river Glubochanka—14 cases of HP, the river Krasnoyarka—14 cases of HP, the river Breksa—13 cases of HP and 1 case of EHP, the river Tikhaya—16 cases of HP.

Surface water quality of the Verkhniy Ertis basin water currents in the period of open water from April to October, 2015 on hydrobiological indicators is non-uniform. According to indicators of perifiton development, it is possible to refer to category of the pure rivers, the river Breksa (background alignment), and the river Bukhtyrma. It is noted, that the highest values of saprobity index are on the river Breksa (below dumpings), river Tikhaya, river Glubochanka and river Krasnoyarka. Other studied waterways were characterized by moderate pollution.

In April-October, 2015 according to macrozoobenthos indicators, the rivers referred to category "pure" are: Bukhtyrma, Kara Ertis, Breksa, Tikhaya (background alignment), Ulba (in the area of Tishinsky minery) and Ulba (background alignment), the river of Glubochanka (background alignment) and the river Oba (background alignment). Less favorable situation has been noted on two points of the river Ertis "for 0.8 km is lower than UK hydroelectric power station dam" and in the river Krasnoyarka "for 1 km is lower than the hollow of the river Beryozovka; at the road bridge", these rivers were characterized by the IVth quality class—"polluted waters". Other waterways were estimated by the IIIrd class of water quality—"moderate polluted".

In surface water of Bukhtyrma and Ust-Kamenogorsk reservoirs, during a research, acute toxicity cases hasn't been revealed, however the small percent of the objects death **test** has been noted. On Ust-Kamenogorsk reservoir the percent of the objects death test varied from 3.3 to 26.6%, and on the Bukhtyrma reservoir mortality of water fleas has made from 3.3 to 43.3%.

By the analysis results on toxicity of the water tests, which were selected on waterways of the Verkhniy Ertis basin in 2015 during 12 months, the following picture was observed: waters of the rivers Kara Ertis, Emel, Ertis, Bukhtyrma, Oba,

Ulba (Ust-Kamenogorsk), Glubochanka (background alignment), Krasnoyarka (background alignment) had no sharp toxic effect on live organisms.

The most unfavourable situation has been noted on the river Ulba (Tishinsky minery). On an alignment cases of acute toxicity "100 m are higher than dumping of mine waters of the Tishinsky minery; 1.25 km are lower than the merge of the river Gromotukhi and Tikhaya" have been registered in February, May, June, August, September and November. The objects death test varied from 63.3 to 100%. On the second alignment "4.8 km are lower than dumping of mine waters of the Tishinsky minery" acute toxicity wasn't observed only in March, during other periods of research concerning water fleas death made from 50 to 100%.

On the river Tikhaya, on an alignment "within the city; 0.1 km higher than the Bezymyannyi stream confluence" the phenomena of acute toxicity have been registered during the second, third and fourth quarter, except December. The objects death test during this period has made from 50 to 100%. On the second alignment only one case of acute toxicity in May has been registered, water fleas death has made 100%.

On the river Breksa, on an alignment "within the city; 0.6 km higher than the mouth of the river Breksa" acute toxicity was observed for the entire period of the research, except March, April, October and December. Water fleas death varied from 80 to 100%. On "background alignment", in June one case of acute toxicity has been registered concerning the objects death test, and it has made 57%.

On the river Glubochanka, on an alignment "0.5 km are lower than dumping of economic-fecal waters treatment facilities (t/f) of the Belousovsky; at the road bridge" during research three cases of acute toxicity were registered: in May, June and August months, objects death test varied from 70% to 90%. One case of the objects death test on "final alignment" has been noted in May, the percent of water fleas death has made 66.7%.

On the river Krasnoyarka, on an alignment "1 km lower than the hollow of the river Beryozovka; at the road bridge" acute toxicity wasn't observed only in February, April, May and December months, during other period objects death test varied from 50% to 100%.

In territory of the EKR extremely high and high pollution are recorded in the following water objects: the river Ulba—27 cases of HP, the river Glubochanka—14 cases of HP, the river Krasnoyarka—14 cases of HP, the river Breksa—13 cases of HP and 1 case of EHP, the river Tikhaya—16 cases of HP.

Information on the actual volumes of dumpings is provided in the table of 16 rivers/according to the Newsletter on state of environment in the Republic of Kazakhstan for 2015.—Astana: RSE "Kazgidromet", 2016 —236 P./(Table 2.9).

The volume of the dumped oil products with sewage in water objects of the EKR has made 0.65132067 thousand tons per year.

The main technogenic impact on natural situation of the area and, including, underground waters, is rendered by mining enterprises, power and metallurgical industry of the Rudny Altai and, generally agricultural production (the livestock production and grain farm, the enterprises processing agricultural production), the industry and separate mining enterprises in valley of the river Ertis and its inflows,

Table 2.9 Pollution of water resources and dumpings of pollutants with sewage

Information on the actual volumes of dumpings		2015 year	2014 year
Industrial dumpings	Volume of water disposal one thousand m^3	54790.776	46059.0172
	Volume of pollutants one thousand tons	41.5	103.25
Economic and household waste waters	Volume of water disposal one thousand m^3	63667.837	67888.8972
	Volume of pollutants one thousand tons	111.828	49.27022
The emergency and not allowed dumpings	Volume of water disposal one thousand m^3	512.2427	25.204
	Volume of pollutants one thousand tons	0.105875	0.0064158
In total (all above-mentioned dumpings)	Volume of water disposal one thousand m^3	118970.8557	113973.1184
	Volume of pollutants one thousand tons	153.433	152.526

for individual cities/Environmental protection and sustainable development of Kazakhstan (2011–2015)/Statistical collection/Committee on statistics of the Ministry of national economy of the Republic of Kazakhstan/

in the east part of the Kazakh hillocky area and intermountain hollows of the Saur-Tarbagatai.

The main sources of the underground waters pollution are:

- mine waters and minery dumps (Shemonaikhinsky, Kamyshin, Nikolaev, Ertis, Belousovsky, Snegirikhinsky, Chekmar, Ridder-Sokolsky, Tishinsky, Maleevsky, Zyryanovsky, Grekhovsky, Ognevsky, Belogorsk, Kendyrlyksky, Akzhalsky MPP, Zhezkentsky minery);
- concentrating factories (Nikolaev, Berezovsky, Belousovsky, Leninogorsk, Zyryanovsky, Ognevsky);
- metallurgical, fuel-energy enterprises (Ertis copper plant, Ertis chemical steel plant, zinc-lead plant, titanium-magnesium plant, Ust-Kamenogorsk and Sogrinsky combined heat and power plant, Leninogorsk zinc-lead plant);
- airport, meat-processing plant, felting and felt plant, other enterprises of the industry in Semipalatinsk (leather and fur enterprise, gas equipment plant, Novoshulbinsky creamery, locomotive depot, meat-processing plant and the dairy shop in the city Ayagoz, Sergiopolsky beer factory);

- livestock enterprises (Volchansky, Shemonaikhinsky, Maleevsky pig factory farms, Torkhanovsky, Leninogorsk, Solovyevsky, Srednegornensky, Pervorossiysky, Yubileinyi, Kamyshin cattle complexes, Cheremshansky, Ust-Kamenogorsk and Komsomol poultry farms, Semipalatinsk poultry farm, cattle complex in Borodulikha, Novobazhenovo, Semenivka, Dmitriyevka, Podnebesnyi, Krivinka, Turksib, Beryozovka, Chugulbay, Mirnyi, Ivanovka, Uruzharsky);
- filtration fields of the cities and other large settlements;
- influence zone of the Semipalatinsk test site.

Polluting components of the specified enterprises are copper, zinc, lead, selenium, manganese, cadmium, ammonia, phenols, and livestock complexes—nitrates and ammonia.

Near Semipalatinsk in the underground waters, the high content of oil products is noted, leading to extremely dangerous extent of pollution, especially if to consider that underground waters have quite close connection with the local basis of unloading here—the river Ertis. Unloading of the underground waters, strongly polluted by the oil products to the river or overflow, to the large water intakes located here, can have catastrophic consequences.

In general, in the area there was a critical situation on underground waters pollution for drinking usage. Auras of pollution have begun to reach such sizes, that in their zone there was 17 of 72 water intakes. In the pollution centers of the JSC "Ulba metallurgical plant", "Ust-Kamenogorsk titanium-magnesium plant", UK MT "Kazzinc" works for the choice of operational wells optimum placement for the polluted stream interception was carried out.

Now, the activities on the polluted stream interception of the pollution center of the JSC "Ust-Kamenogorsk titanium-magnesium plant" are being carried out, and they have begun since 2005 in the territory of the JSC "Ulba metallurgical plant" tail economy. The activities aimed at the polluted stream interception in territory of an industrial site and dump economy of the UK MT "Kazzinc" enterprise are practically not carried out. Reconstruction of drainage water intake, which was recommended in 1997 hasn't been executed (The program of development of the East Kazakhstan Region territory on 2016–2020; Environmental protection and sustainable development of Kazakhstan 2011–2015).

For the decrease in the extent of underground waters pollution, it is necessary to make:

- the analysis of water consumption balance and water disposal of the enterprises for justification of use or drainage waters dumping at the polluted streams interception by reduction of drinking water consumption for the technical purposes, water recirculation development, termination of dumping into the city sewerage of conditionally pure and intensively polluted because of metals, oil products industrial drains without local cleaning;
- the annual assessment and holding analysis of the water preserving actions are obligatory, if necessary recommendations about their improvement are proved;

- the careful inspection of protective waterproofing concerning water bearing communications of reverse water supply and technological knots on sites of pollution auras;
- the reduction of the underground waters pollution concerning alluvial deposits due to decrease of the polluting components carried out from the burial grounds and industrial dumps by their recultivation. The intensity of the polluting components carrying out at almost unlimited number of the easily saved-up soluble waste is defined by the intensity of food infiltration with atmospheric precipitation. The exception of atmospheric food will allow "to preserve" waste;
- the construction of reliable barriers in the closing pollution aura alignment with the trial pumping organization of the polluted underground waters and their cleaning before maximum-permissible dumpings.

Additional works are necessary for the justification of drainage water intakes and drains utilization in order to decrease in extent of pollution in the centers such as the Tishinsky minery (Ridder city) and Zyryanovsky tailings dam.

The main components, polluting underground waters are: cadmium, thallium, oil products, manganese, cyanides, selenium, iron, ammonia, lead.

Macrogeosystems integration in the river basin Ertis is caused by the lateral movement of the substance, subordinated to the general process of geographical drain from the nival-glacial zone of drain formation to the delta. Each rank of natural complexes, concerning topological dimension, has certain fluctuation limits of substance amounts in components. To all ranks of geosystems corresponds concrete landscape ranks with substance reserves, which quantitative parameters define extent of substance migration in the conditions of moistening deficiency and depend on the drain module. The aforesaid, explains the spatial differentiation of landscapes concerning transit zones, feathering out, the drain dispersion and the intensity of their functioning, depending on the intensity of the chemicals' movement, technogenesis loading.

Soils of the region are the subject to technogenic pollution. From them, the most fully technogenic pollution of soils by heavy metals is studied in the Kazakh part of the Big Altai. Here mainly technogenic abnormal field is mapped in the central and northwest parts of region, where the main industrial facilities are localized, making the negative impact on the environment.

The soil cover of the city is a difficult and non-uniform natural and anthropogenous biogeochemical system. In the background of artificial technogenic formations such as asphalted streets, highway areas, there are anthropogenously changed and natural soils (yards, parks boulevards, and waste grounds). Products of the technogenesis drop out on the land surface and are collected on the top horizons of soils, changing their chemical composition and joining again the natural and technogenic cycles of migration. By the geochemical character change of the natural and poorly modified city soils of the East Kazakhstan, in reference to the background soils of the region, it is possible to judge about the extent of their technogenic transformation (Fortescue 1985; Moiseenko 1989; Felenberg 1997; Protasov 2000).

The essential value for the formation of the geochemical background of the East Kazakhstan cities' soils have the duration and industrial development nature of the city in historical time, i.e. since the XVIII century, when the first mines began to be developed.

According to the impact effect on city soils, the technogenic substances can be united in two groups. Pedogeochemically active substances prevail in weight emissions, and change alkaline-acid and oxidation-reduction conditions in soils. These are generally nontoxic and slightly toxic elements with high percentage abundance—iron, calcium, magnesium, alkalis and mineral acids. At the achievement of certain limit, acidulation or alkalifying affects soil flora and fauna. Some gases are also pedogeochemically active, for example, the hydrogen sulfide and methane, changing an oxidation-reduction situation of migration. Biochemically active substances affect, first of all, live organisms. These are usually typomorphic for each production type, which are highly toxic pollutant with low percentage abundance (mercury, cadmium, lead, antimony, selenium, and etc.) forming auras, more contrasting concerning the background, and constituting danger to biota and the person.

In the cities, intake of dust on surface is much more, than in the natural background landscapes. Iron, calcium and magnesium prevail in city dust macrocells. Two geochemical consequences of athmotecnogenic dust supply on the city territory are connected with the ferruginization of soils which is almost not influencing alkaline-acid and oxidation-reduction conditions of elements migration and the carbonatization of soils leading to the increase in their alkalinity, saturation of the absorbing complex bases, binding of many metals in almost insoluble carbonates. At considerable and long intake of carbonate dust to sour and neutral soils, there is a change of landscape water migration class. Sour, sour gley $(H^+, II^+ - Fe^{2+})$, neutral and neutral gley $(H^+ - Ca^{2+}, H^+ - Ca^{2+}, Fe^{2+})$, classes are transformed in calcic and calcic gley $(Ca^{2+}, Ca^{2+} - Fe^{2+})$, classes of water migration.

Alkaline technogenic transformation of city soils leads to change of their buffer action, increase in absorbing ability, reduction of carrying out possibility and migratory ability of many pollutant and first of all, heavy metals. In steppe and desert zone processes of soils carbonatization are less noticeable. Pollution of city soils by macro and micro-elements is followed by the transformation of soil and geochemical structure of the territory. First of all, radial geochemical differentiation of soil profile due to pollutant accumulation in the top horizons sharply increases. In podsolic and gray forest soils, technogenic accumulation shades the eluvial and illuvial background profile differentiation. On the contrary, in black earth uniform distribution of metals is replaced by the superficial and accumulative one. Athmotecnogenic pollution of autonomous soils, strengthening of storm superficial drain, flooding by the polluted ground waters defines the accumulation of toxic substances in the subordinated landscape soils. In this regard, the interfaced types of soil and geochemical catenas prevailing in the cities background conditions are replaced by the sharply differentiated accumulative types. The unevenness of soil cover pollution of the cities leads to the casual ratios emergence of chemical elements between soils of the autonomous and subordinated landscapes.

Detailed researches of city "relief" influence have shown that the residential industrial buildings serve as mechanical barriers for air migrations of the technogenic substances, and near them in soils more contrast anomalies of pollutant are formed. Concentration of the industry, power system, automobile transport and municipal waste in the cities leads to the formation of heavy metals and other minerals of technogenic anomalies in the city soils.

The prevailing impact on land resources condition of the East Kazakhstan region is made by the enterprises of agriculture, mining industry, and power system.

On the basis of data, submitted by nature users of SI Ecology Department in the EKR, works on registration and accounting of pollution sites are carried out. At the moment, in total 282 pollution sites are registered across the East Kazakhstan region.

For the spring period in Ust-Kamenogorsk, excess of MPC on concentration of heavy metals has been recorded in the following districts of the city:

- at the crossing of Traktornaya street and Abay prospectus (1 km to south-east from an industrial site of JSC "Kazzinc") the concentration of cadmium is 9.0 MPC, lead—5.1 MPC, copper—1.8 MPC, zinc—1.6 MPC;
- at the crossing of Rabochaya and Bazhova streets (1 km from JSC "Kazzinc") the concentration of cadmium makes 31.4 MPC, copper—21.2 MPC, lead—14.0 MPC, zinc—13.3 MPC;
- near the highway of Lenin avenue (the area of RSU, 3 km to south-west from JSC "Kazzinc") the concentration of cadmium—11.2 MPC, lead—4.6 MPC, zinc—1.2 MPC;
- near the "Blue Lakes" park (3 km from JSC "Kazzinc") the concentration of copper is 2.2 MPC, zinc–1.6 MPC, cadmium—1.2 MPC;
- in the territory of № 34 school (3 km from JSC "Kazzinc") the concentration of lead makes 3.5 MPC, cadmium—3.2 MPC.

In soil tests the content of chrome was in normal limits.

For the autumn period in Ust-Kamenogorsk, the excess of MPC on concentration of heavy metals has been recorded in the following districts of the city:

- at the crossing of Traktornaya street and Abay prospectus (1 km to south-east from an industrial site of JSC "Kazzinc") the concentration of cadmium—10.2 MPC, lead—2.5 MPC, copper—1.5 MPC, zinc—2.1 MPC;
- at the crossing of Rabochaya and Bazhova streets (1 km from JSC "Kazzinc") the concentration of cadmium—14.2 MPC, copper—6.2 MPC, lead—7.1 MPC, zinc—9.6 MPC;
- near the highway of Lenin avenue (the area of RSU, 3 km to south-west from JSC "Kazzinc") the concentration of cadmium is 5.3 MPC, lead—4.3 MPC, zinc—1.8 MPC, copper—1.5 MPC;

– near the "Blue Lakes" park (3 km from JSC "Kazzinc") the concentration of cadmium—2.6 MPC, lead—1.9 MPC, zinc—1.1 MPC;
– in the territory of № 34 school (3 km from JSC "Kazzinc") the concentration of cadmium—6.4 MPC, lead—4.2 MPC.

In soil tests the content of chrome was in normal limits.

For the spring period in Ridder, excess of MPC on the concentration of heavy metals has been recorded in the following districts of the city:

– around park zone the concentration of cadmium—12.6 MPC, lead—10.6 MPC;
– near the sanitary-protection zone of Zinc plant, the concentration of cadmium has made 9.8 MPC, lead—9.6 MPC, copper—2.2 MPC, zinc—1.2 MPC;
– near the sanitary-protection zone of Lead plant, the concentration of cadmium—25.0 MPC, lead—11.7 MPC, copper—2.9 MPC, zinc—1.9 MPC;
– around the territory of № 3 school, concentration of cadmium—29.0 MPC, lead —11.7 MPC, copper—2.8 MPC, zinc—1.7 MPC;
– near the most busy highway, concentration of lead is 8.8 MPC, cadmium—1.2 MPC.

In soil tests the content of chrome was in normal limits.

For the autumn period in Ridder, excess of MPC on concentration of heavy metals has been recorded in the following districts of the city:

– around park zone concentration of cadmium—2.5 MPC, lead—4.1 MPC;
– near the sanitary-protection zone of Zinc plant, where concentration of cadmium has made 9.2 MPC, lead—1.3 MPC, copper—9.6 MPC, zinc—1.4 MPC;
– near the sanitary-protection zone of Lead plant, where the concentration of cadmium comprises 15.1 MPC, lead—21.5 MPC, copper—11.8 MPC, zinc — 3.1 MPC;
– around the territory of № 3 school, the concentration of cadmium—6.0 MPC, lead—6.4 MPC, zinc—1.6 MPC;
– near the most busy highway, concentration of lead—7.6 MPC, cadmium—5.6 MPC, zinc—4.3 MPC, copper—4.2 MPC.

In soil tests the content of chrome was in normal limits.

For the spring period in Semey around the № 3 school and in the highway territory of Kabanbai batyr street, the concentration of copper respectively exceeded norm—5.4 and 5.0 MPC.

For the autumn period in Semey, chrome concentration were in limits of 0.002–0.3 MPC, cadmium 0.1–0.5 MPC, zinc—0.4–0.9 MPC, lead—0.4–0.6 MPC and copper 0.2–3.5 MPC. Near the sanitary-protection zone of "Semeycement", the concentration of copper have exceeded the standard—3.5 MPC.

In the region as of 01.01.2016 year, 161 subsoil users, and 13 water users are registered.

In gold mining, there are 9 foreign and 15 domestic companies, which have contracts. The large ones among them, respectively are: the JSC FIC "Alel" (financial and investment corporation), the JSC "Charaltyn", LLP "Kazzinc",

LLP OMC "Andas-Altyn", JSC "Semgeo", LLP "Toskara". JSC "Kazzinc" and LLP "Vostoktsvetmet", SLLP "Ore mining enterprise "Sekisovskoe" company of "Hambledon Mining Company Limited" are the largest ones in the subsurface management area on polymetallic ore. LLP «Satpaevsk Titanium Mines LTD» (LLP «STM») is engaged in extraction of ilmenite at working off of the Satpayevsky field; the JSC "Ulba metallurgical plant" extracts fluorite on the Karadzhalsky field.

Solid combustible minerals are extracted by four subsoil users, from them the largest one is LLP "Karazhyra LTD".

Domestic company the LLP "TEMP" (Temirtau electric metallurgical plant) is engaged in extraction of ferrous metals (manganese).

The bulk of subsoil users are engaged in popular minerals extraction: these are bentonite clays, volcanic tufa, gabbro, clays, granites, diorites, limestones, quartz sand, ceramsite clays, brick clays, sand-gravel mixes, table salt, porfirita, and construction stone.

In the area, also production of mineral water by three subsoil users is conducted: LLP "Zaysan sulary", SP Churkumbayev M.S., LLP "Rakhmanovskiye klyuchi".

Part of subsoil users in 2015 didn't carry out production activity for various reasons (LLP "Arman", LLP "Semgeo", LLP "Zherek", LLP Ore mining company "Andas Altyn", and etc.).

During work on mining and construction of subsurface use facilities, removal and preservation of fertile layer is made for the subsequent use for land reclamation, according to requirements of the RK Ecological code 220 articles. The need for removal and storage concerning soil and fertile layer is defined in materials of the soil research territories, conducted at withdrawal of the land plots [163–170].

For the purpose of decrease in the volumes, placed in environment of the mining subsoil users waste in region, they use the overburden and containing breeds for filling of fulfilled pit space. The LLP "Karazhyra LTD", SLLP "OME "Sekisovskoye" of the "Hambledon Mining Company Limited" (to decipher), LLP "Vostoktsvetmet", LLP "Kazzinc" of RMPE, ZMPE use for layings in the fulfilled mine development production wastes.

The systematic work on filling of the fulfilled pit part with overburden breeds is carried out by LLP "Karazhyra LTD", which realizes coal mining on the homonymy field. Specified enterprise distributes the overburden breeds, forming by production of mining operations to the fulfilled pit space (an internal dump), thereby the technical stage of land reclamation has been accomplished.

According to the qualification of the Ministry of environmental protection of the RK, on pollution of soils, underground and surface water, vegetation, these settlements such as the Glubokoe, Belousovka, Verkhneberezovsky, Cheremshanka, and Uvarov are possible to refer to ecological catastrophe zone.

Near the Ertis river course, the intensity of soils of technogenic pollution by 2–3 times exceeds the background. In the Pervomay settlement, soils pollution by 4–32 times exceeds the background, and in the Ust-Talovka settlement—by 16–32 times. In geosystems of lower current of the river Naryn, there was an observation that soils pollution by 2–16 times exceeds the background.

Radiation condition. First of all, this is the influence of SNTS and enterprises of nuclear industry complex in Abralinsky district of the former Semipalatinsk region and part of territories of the Pavlodar and Karaganda regions; its area makes about 618 thousand km^2. Here 470 nuclear explosions, from them 26 is overland, 87 is air and 357 is underground, including the first Soviet nuclear explosion with power of 20 kg plutonium bomb were conducted (on August 29, 1949 year). SNTS was closed by the Decree № 409 of the President of Kazakhstan in August 29, 1991 year. The SNTS occupies the space of 18,500 km^2.

Considering the improvement of the environment quality, which was the result of production declining, and taking data on the total quantity of pollutants emission and drain for 1996–1998, it is possible to tell that for the stabilization of environment quality, it is necessary to limit the total quantity of pollutant emissions from the stationary sources of pollution within 250–300 thousand tons per year. But in this assessment, qualitative composition of the thrown-out substances isn't considered.

The complex mineralogical composition of the raw materials, processed at the enterprises of nonferrous metallurgy and low maintenance, in their useful components cause the biggest specific exit of waste in the extracting branches. The costs of their transportation and warehousing often exceed 40% of valuable components extraction cost. In dumps of production chain "production—enrichment-metallurgy" more than a third of non-ferrous and precious metals extracted with ores is lost, and huge reserves of various potential construction raw materials are frozen.

For many years of the enterprises functioning concerning mineral-raw complex in region, the huge number of enrichment waste and metallurgical conversion of ores has been accumulated.

The municipal solid waste is the heterogeneous mixes, which are differing on qualities and fineness, and formed in the process of people activity. Many of them drop out or destroyed as useless or undesirable. Their basis consists of waste paper (20–40% on weight), food waste (25–40%), textiles (4–6%), glasses (4–6%), ferrous and non-ferrous metals (2–5%), rubber (2–4%), polymers (1–2%), and etc. The general content of organic substances in waste fluctuates from 50 to 80% counting on absolutely solid, and humidity fluctuates depending on season of year within 45–60%. Besides, there is a seasonal change in morphological composition of municipal solid waste (Bazarbayev et al. 2002; Geosistemnyiy monitoring 1986; Glazovskaya 1988).

The enrichment waste and ores processing have lately begun to be used strenuously in construction repair work not only in the Ertis river basin, but also outside Kazakhstan. Especially, widely specified raw materials find application in the cities of Ridder, Zyryanovsk and their vicinities. So, for example, "the easy fraction"—waste of polymetallic ores enrichment in heavy suspensions of Tishinsky and Zyryanovsk concentrating factories—goes for highways dumping and repair, railway embankments, access roads, finds application in individual construction of houses, garages, garden lodges and many other. Chashinsky storage tails in Ridder city are used for preparation of plaster solution, clinker is used to "improve" a roadbed in the cities, on again under construction routes. Levels of heavy metals concentration—Pb, Cu, Zn, Ag, Ba, Mo, Sn, Cd, Sb—in widely used MB (main base) are rather high, that promotes intensive pollution of soils in roadside zones and atmosphere secondary pollution.

Principle solution of MSW problem concludes in development of low-waste and waste-free industrial technologies—the basic principle of resource use, in effective and complex use of natural resources, drawing into economic circulation of production wastes and consumption.

For MMF (man-made mineral formation) the positive forecast will be connected with the consecutive reduction of dump product volumes that can be reached as directly by elimination of the saved-up dump products (by their repeated processing), and due to the reduction of new released volumes of waste (and the related toxic substances) at the production and processing of mineral raw materials. It is possible only at the introduction of complex low-waste technologies of mineral raw materials processing.

Toxic substances pollution of atmosphere comes from MMF due to dusting of waste stores: dumps of off-balance ores, slags, breeds, and drained tailing dams. Considering large reserves of mineral raw materials in East Kazakhstan, the essential sources of pollution are the opened ones, but not used in fields. Dusting from the destruction of all technogenic waste in the East Kazakhstan makes about 111 thousand tons per year. Dust contains such toxic components as lead, zinc, copper, cadmium, mercury, selenium, tellurium, arsenic, tin, and etc. There is no representative data on the distribution and stay forms in stores of dump products concerning toxic elements. Moreover, the available data (passport of waste) on the structure of dump products in essence don't reflect the current real state, since these data characterize the general structure of dump products at the time of accumulation in stores. Considering that most storages of waste exist decades, it is competent to expect very essential changes in the distribution and forms of finding dump product components. So, for example, in tailing dams and in dumps of off-balance polymetallic ores quite intensive electrochemical reactions promoting the formation of soluble connections in number of elements, including toxic proceed. Also long (decades) safety in a liquid phase of the tailings dam of cyanides was found. However, such researches have sporadic character and, in this regard, now it is impossible to determine the conditions of toxic substances migration from the stores in the environment components, and the areas of their dispersion and concentration.

The total waste stocks of mining and metallurgical production of non-ferrous, rare, precious and radioactive metals are already comparable to stocks corresponding rather large-scale deposits. Along with the main useful components at dump products, there are numerous element satellites, in some cases being of independent value. At the same time, both among the main, and accompanying components there are toxic elements (and their connections) in significant amounts such as lead, zinc, cadmium, mercury, arsenic, antimony, selenium, sulfur, and etc. The last in processes of production and processing of ores collect in storages, which goes to the environment (soil, water, air, and biota). Available archival and literature data, and also results of control approbation components, concerning the environment, show that the soils and waters contain complex of toxic components in the quantities considerably, exceeding near the stores of production waste and processing of polymetallic ore, MPC. So, the content of arsenic, cadmium, lead, zinc, mercury, etc. in some cases exceeds MPC by 3–5 times. For the waste stores, taken at production and processing of gold ores, the main pollutant of water and soils are arsenic and tin, which content is quite often exceeded here by MPC in ten times. Besides, here usually there are mercury and cyanides (Weyant 1997; Mage and Zali 1992; Vorobeychek 1994).

In influence zones of waste storages in the rare metal productions as sources of radiation hazard, the main pollutants of soils and waters are low-toxic elements: lithium (LI); beryllium (Be); rubidium; (Rb); caesium (Cs); thallium (TI); niobium (Nb); cobalt (Co). Due to presence of radioactive isotopes ^{22}Ra, ^{137}Cs and ^{60}Co in the influence zones of rare metal stores is distinctly shown the increased radioactivity (gamma rays ranging from 60 to 500 mcR/h). The sharply increased radioactivity is noted in the influence zones of waste stores in the productions, which are connected with processing of radioactive ores (to 5000 and more than mcR/h). In the region territory, except registered 27 storages of radioactive waste, there are numerous points of raised radiation connected with various technogenic sources. In the territory of Ust-Kamenogorsk, there are about 400 anomalies studied in detail for only 12%.

The main large enterprises of nonferrous and ferrous metallurgy of the Northeast industrial region, which activity depends on water resources of the Ertis river are listed below:

- Ust-Kamenogorsk lead—zinc plant;
- Ridder lead—zinc plant;
- Zyryanovsky lead plant;
- Ertis polymetallic plant;
- East Kazakhstan copper—chemical combine;
- Zhezkent mining and processing plant;
- JSC "Ispat-Karmet";
- JSC "Kazzinc";
- JSC "Ust-Kamenogorsk titanium and magnesium plant".

The main characteristics of some plants.

1. Ust-Kamenogorsk lead—zinc plant

Productive power:

- Electrolytic zinc—186,400 tons per year (total amount in the Republic of Kazakhstan 292,900 tons per year);
- Purified lead—145,900 tons per year (total amount in the Republic of Kazakhstan 326,100 tons per year);
- Electrolytic copper—40,000 tons per year.

The number of employees is over 9000 people.

Raw materials are bought from Yubileinyi - Snegirikhinsky minery and others in the Republic of Kazakhstan.

The plant is complex, color and metallurgical, and has melting factories of lead, zinc, copper and other precious metals and produces such metals as zinc, lead, copper, gold, silver and cadmium, both the synthetic and processed products from zinc and lead, also includes the Ertis copper-chemical combine.

2. Ridder lead—zinc plant

The plant is located on the river Ulba at the distance of 80 km from Ust-Kamenogorsk in the northeast direction and engaged in thorough production from minery before receiving the cleared products. It consists of 4 mineries: Tishinsky, Ridder, Lenin and Shubinsky mineries; 2 concentrating factories (5.4 million tons per year); the lead-melting factory on production of lead on the basis of accumulator battery skrap processing. The rights of economy management are transferred to the "Ridder Investment". As the final products, it produces such metals as zinc, copper, lead, and also precious metals, sulfuric acid, cadmium, zinc-aluminum alloys, antimoni-lead alloys and other metal products.

3. Zhezkent mining and processing plant

Among 6 plants of the East Kazakhstan region, it is the only one located in territory of the former Semipalatinsk region, but, nevertheless, it is located close to border with the former East Kazakhstan region. Moreover, it is close to the border with the Russian Federation. Also, it includes underground minery, concentrating factory under construction, mechanical-repair manufactory, pit on stowage material extraction and other enterprises. The main industrial center is the settlement of Zhezkent city type.

4. Zyryanovsky lead plant

This plant is one of the largest mining enterprises in the Republic of Kazakhstan and is located in Zyryanovsk (population—51,500 people), at the distance of 160 km in the southeast direction from Ust-Kamenogorsk. It has three mineries: Zyryanova, Grekhovsk and Maleevsk, and concentrating factories. Enriched ores almost completely are delivered in the melting plants of Ust-Kamenogorsk.

5. Ertis polymetallic plant

The plant is located in the northwest district of Ust-Kamenogorsk. It consists of Belousovky minery and concentrating factory (at the distance of 25 km in the northern direction from Ust-Kamenogorsk), the Ertis minery (at the distance of 12 km in the east direction from Berezovsky) and Berezovsky concentrating factory (at the distance of 55 km in the northwest direction from Ust-Kamenogorsk). All these enterprises are located close to the river Ertis.

6. East Kazakhstan copper—chemical combination

The plant had three mineries of the open-cast mountain mining Nikolaev, Shemonaikhy, Kamushinsky, however, in 1994 it completed its production in Kamushinsky minery and, now, the plant is extracting ores in two mineries. Ores are difficult polymetallic, enriched at the Nikolaev concentrating factory. The plant produces enriched copper, zinc ores and mixed copper-zinc ores.

Each of these plants makes contribution to water pollution of the Verkhniy Ertis river basin. At the same time, the most characteristic pollutants are as following: copper, zinc, lead, aluminum, mercury, and oil products. Besides, a large amount of pollutants arrives from the cities of Ust-Kamenogorsk and Ridder together with sewage.

The greatest anthropogenous load on water resources of the Ertis river basin is the share of riverheads within the East Kazakhstan region. Here the largest enterprises of nonferrous metallurgy of the Republic of Kazakhstan are concentrated, which are known to be the main pollutants of the Ertis river and its inflows.

The characteristic feature of all these enterprises is the low level of use concerning systems reverse water supply.

Copper, zinc, lead, oil products, phenols, and etc., dumped with sewage in water objects can be considered as the main pollutants.

In the initial section of the river Ertis within the territory of RK, there are no direct dumpings from industrial enterprises. Hydrochemical mode of the river Kara Ertis in alignment of the Buran village is formed due to washing away and dissolution of rocks, superficial drain from reservoir in the territory of the PRC, but a certain impact on water quality is exerted by the polluted superficial drain, and dumpings from agricultural fields in the territory of the PRC. In waters of the river Kara Ertis, arriving from the territory of the PRC, the concentration of 1–2 MPC contain oil products and nitrates, besides, there are copper, zinc and other elements.

In recent years, some depression of the general impurity level and maintenance of separate pollutant elements becomes perceptible, however there are cases of extremely high level of water pollution in the river Ertis, mainly by copper and zinc (to 100 MPC). In intra annual change of water quality connected with the accurate regularities has not been revealed and impurity level augmentation during separate seasons and months are bound mainly to the volley dumping of sewage.

Technogenic factors operating on geosystems of the East Kazakhstan influence also the human beings. Considering population ageing at deterioration of ecological situation, it is necessary to expect further augmentation of disease rate, disability and mortality from malignant neoplasm, especially from lung cancer, trachea and bronchus among men.

According to EKRTDEP, about 95% of the Kazakhstani populations in Big Altai live in the territories, which are in varying degree ecologically unsuccessful and, about 70% of inhabitants are concentrated in the settlements situated in the centers of industrial production, the cities qualified as zones of ecological catastrophe.

According to the data collected above, there was defined an ecological situation in the East Kazakhstan and possible development of the situation in the future as a general condition image.

Thus, for the last decades in the Ertis macrosystem, especially in its Rudny Altai part, there was a glut of the industrial enterprise territory, particularly the productions of mineral and raw complex. The last ones emit more than a half of the harmful substances coming to the environment. The largest enterprises, known as the pollutants of the Altai environment, are localized in the following subgeosystems: the Actually-Shulbinsky, Ubinsky, Ulbinsky, Kyzylsu-Taitinsky, Bukhtyrminsky, and etc. On average, for each resident of the Rudny Altai region, there are over 250 kg emissions of harmful substances per year.

Environmental degradation, social and economic destabilization has led to the decrease in the living standard and unstable level of human development. Therefore, improvement of an ecological situation, creation of more favorable conditions for people's life and work is considered to be the major and urgent action for the environmental protection. Here the population mortality in comparison with rural areas has almost doubled (670–1100 for 100,000 population).

In the conditions of Kazakhstani economy transition to sustainable development, the problem of ensuring the economic progress, not breaking the balance of natural geosystems, is particularly acute. In respect of strategic tasks, for the further development of the region, it is necessary to develop a long-term environmental policy. In its turn, for this purpose the creation of theoretical base, development of the methodological principles, their coordination with practice requirements and region conditions is demanded.

References

Alekseenko VA (1990) Geochemistry of the landscape and the environment. Science, Moscow, p 140

Armand DL (1975) Teoriya polya i problema vyideleniya geosistem (Theory of field and problem of geosystems allocation), Geography issues, pp 29–30

Aslanikashvili AF, Saushkin BG (1975) Novyie podhodyi k resheniyu metodologichesikh problem sovremennoy geograficheskoy nauki (New approaches to the methodological decision on problems of the modern geographical science). Geography in the Georgian SSR. Nauka, Tbilisi, pp 15–51

Bayandinova SM (2003a) K voprosu problem tehnogennogo zagryazneniya gornyih geosistem Yugo-Zapadnogo Altaya (To problems issue of technogenic pollution on mountain geosystems of the Southwest Altai). Al-farabi Kazakh National University, Almaty, pp 167–172

Bayandinova SM (2003b) Analiz tehnogennogo zagryazneniya prirodnoy sredyi Kazahstanskogo Altaya (Analysis of technogenic environment pollution of the Kazakhstan Altai). Reporter of Al-farabi Kazakh National University. № 1 (12), Ecology series, pp 1–14

Bayandinova SM (2003c) Osobennosti geoekologicheskoy obstanovki prirodnoy sredyi Vostochnogo Kazahstana (Features of geoecological situation on environment of the East

Kazakhstan). Reporter of Al-farabi Kazakh National University № 2 (17). Geography series, pp 17–23

Bayandinova SM (2003d) Sovremennyie tehnologicheskie podhodyi v izuchenii tehnogennogo zagryazneniya prirodnoy sredyi Yugo-Zapadnogo Altaya (The modern technological approaches in study of technogenic environment pollution of the Southwest Altai). EKSU, Ust-Kamenogorsk: Novosibirsk-Almaty, pp 116–119

Bayandinova SM (2003e) Faktoryi tehnogennogo zagryazneniya geosistem Vostochnogo Kazahstana (Factors of geosystems technogenic pollution of the East Kazakhstan). KSU named after E.A. Buketov, Karaganda pp 22–26

Bayandinova SM (2005a) Geosistemno - basseynovyiy podhod v izuchenii prirodnoy sredyi Vostochnogo Kazahstana (Geosystem- pool-type approach in environment study of the East Kazakhstan). Al-farabi Kazakh National University, Almaty, pp 54–59

Bayandinova SM (2005b) Analiz tehnogennogo zagryazneniya prirodnoy sredyi Vostochnogo Kazahstana (Analysis of environment technogenic pollution of the East Kazakhstan). Reporter of Al-farabi Kazakh National University № 2 (21), Geography series. pp 160–164

Bazarbayev SK, Burlibayev MZ, Kudekov TK, Murtazin EZ (eds) (2002) Sovremennoe sostoyanie zagryazneniya osnovnyih vodotokov Kazahstana ionami tyazhelyih metallov (Current state of the main water currents pollution in Kazakhstan ions of heavy metals) (2002) Khaganate, Almaty, p 256

Bokov VA (1977) Fiziko-geograficheskie granitsyi i vyidelenie geosistem (Physiographic boundaries and separation of geosystems). Problems of natural zoning. Nauka, Ufa, pp 332–347

Bykov BA (1988) Ekologicheskiy slovar (Ecological dictionary). Nauka, Alma-ata, p 714

Cherednichenko AV, Nedovesov VS (1997) K otsenke vklada vyibrosov otdelnogo predpriyatiya v prizemnyie kontsentratsii vrednyih veschestv (To assessment of emissions contribution on the separate enterprise in ground concentration of harmful substances). Al-farabi Kazakh National University, Almaty, pp 42–44

Cherednichenko AV (2002) Osnovnyie istochniki vyibrosov i uroven zagryazneniya vozdushnogo basseyna Kazahstanskogo Altaya (Main sources of emissions and level of air basin pollution of the Kazakhstan Altai). Reporter of Al-farabi Kazakh National University № 1 (10). Ecology series, pp 31–34

Chigarkin AV (2003) Geoekologiya i ohrana prirodyi Kazahstana (Geoecology and nature conservation of Kazakhstan). Kazakh Universities, Almaty, pp 41–55

Demek Ya (1977) Teoriya sistem i izucheniya landshafta (Systems theory and landscape study). Progress, Moscow, p 336

Dergunov YA, Krasov VI, Melyaev VB (1988) Issledovanie dinamiki promyishlennyih vyibrosov v atmosferu goroda (Research on dynamics of industrial atmosphere emissions in the city). Hydrometeorological Publication, pp 76–85

Dreyer OK (1997) Ekologiya i ustoychivost razvitiya (Ecology and stability of development) RIEU, Moscow, pp 16–35

Dyakonov KN (1985) Metodologicheskie problemyi izucheniya fiziko-geograficheskoy differentsiatsii (Methodological problems of studying on physiographic differentiation). Geogr Issues 13–20

Dyakonov N (1988) Prostranstvenno-vremennaya sopryazhennost prirodnyih protsessov na globalnom i regionalnom urovne. Globalnyie problemyi sovremennosti i kompleksnoe zemlevodenie (Spatio-temporal conjugacy of natural processes at global and regional level. Global problems of present and complex physical geography). Nauka, Moscow, pp 112–116

Dzhanaleeva GM (1986) Antropogennyie modifikatsii landshaftov delt rek i puti ih ratsionalnogo ispolzovaniya. Sb.: Prirodno-resursnyiy potentsial Vostochnoy Sibiri i problemyi formirova-niya agrarnyih i promyishlennyih kompleksov (Anthropogenous modifications of landscapes on river deltas and way of their rational use. Coll.: Natural and resource capacity of the East Siberia and problem of formation on agrarian and industrial complexes). Irkutsk: Geography Institute SB SA of SSR, pp 54–67

Dzhanaleeva GM (1993) Geosistemnyiy podhod v izuchenii landshaftov Balhash-Iliyskogo regiona (Geosystem approach in landscapes study of the Balkhash-Iliysky region). Dissertation work of geography science researcher. Almaty, p 356

Dzhanaleeva KM (ed) (1998) Fizicheskaya geografiya Respubliki Kazahstan (Physical geography of the Republic of Kazakhstan). Kazakh Universities, Almaty, p 319

Dzhanaleeva GM, Bayandinova SM (2004) Teoreticheskie osnovyi i metodologicheskie problemyi geosistemno–basseynovogo podhoda k izucheniyu prirodnoy sredyi. Sb. materialov mezhdunarodnoy nauchno-prakticheskoy konferentsii «Teoreticheskie i prikladnyie problemyi geografii na rubezhe stoletiy» (Theoretical bases and methodological problems of geosystem-pool-type approach to environment study. Coll. materials of the international scientific and practical conference "Theoretical and application-oriented problems of geography at turn of centuries"). Arkas, Almaty, pp 28–31

Dzhanaleeva GM, Bayandinova SM (2003) Geoekologicheskie problemyi ispolzovaniya mineralno-syirevoy bazyi Kazahstanskoy chasti Bolshogo Altaya (Geoecological problem of mineral resources use in the Kazakhstan part of Big Altai). Reporter of Al-farabi Kazakh National University № 1-2,. Geography series, pp 101–105

Environmental protection and sustainable development of Kazakhstan (2011–2015) Statistical collection. Committee on statistics of the Ministry of national economy of the Republic of Kazakhstan

Felenberg P (1997) Zagryaznenie prirodnoy sredyi. Vvedenie v ekologicheskuyu himiyu (Pollution of environment. Introduction to ecological chemistry). Mir, Moscow, pp 6–15

Fortescue D (1985) Geohimiya okruzhayuschey sredyi (Environment geochemistry). MSU, Moscow, p 179

Garmashova SA, Cherednichenko AV (2002) O vliyanii vyibroosov Leninogorskoy TETs na okruzhayuschuyu sredu (About influence of emissions on the Leninogorsk combined heat and power plant on the environment). Reporter of Al-farabi Kazakh National University № 1 (10), Ecology series, pp 38–40

Geograficheskoe prognozirovanie priroohrannyih problem (Geographical forecasting of nature safety problems) (1988) MSU, Moscow, pp 58–62, 86–96

Geohimiya tyazhelyih metallov v prirodnyih i tehnogennyih landshaftah (Geochemistry of heavy metals in natural and technogenic landscapes) (1983) MSU, Moscow, p 117

Geohimiya prirodnyih i tehnogennyih landshaftov SSSR (Geochemistry of natural and technogenic landscapes of the USSR) (1988) Vysshaya Shkola, Moscow, p 419

Geosistemnyiy monitoring. Stroenie i funktsionirovanie geosistem (Geosystem monitoring. Structure and functioning of geosystems) (1986) GI of Academy of Sciences of the USSR, Moscow, p 355

Glazovskaya MA (1962) O geohimicheskih printsipah klassifikatsii prirodnyih landshaftov. Geohimiya stepey i pustyin (About geochemical principles of classification on natural landscapes. Geochemistry of steppes and deserts). MSU, Moscow, pp 22–39

Glazovskaya MA (1964) Geohimicheskie osnovyi tipologii i metodiki issledovaniy prirodnyih landshaftov (Geochemical fundamentals of typology and researches technique of natural landscapes). MSU, Moscow, p 611

Glazovskaya MA (1967) Landshaftno-geohimicheskoe rayonirovanie sushi Zemli (Landscape and geochemical division of the Earth land). Reporter of MSU № 5. Geography series, pp 6–23

Glazovskaya MA (1976) Aktualnyie problemyi teorii i praktiki geohimii landshaftov (Urgent problems on theory and practice of landscapes geochemistry). № 2. Reporter of MSU. Geography series, pp 46–54

Glazovskaya MA (1988) Geohimiya prirodnyih i tehnogennyih landshaftov (Geochemistry of natural and technogenic landscapes). M.: Vysshaya shkola, p 286

Glazovskaya MA (1992) Biogeohimicheskaya organizovannost ekologicheskogo prostranstva v prirodnyih i antropogennyih landshaftah kak kriteriy ih ustoychivosti (Biogeochemical organization of ecological space in natural and anthropogenous landscapes as criterion of their stability). Notion of RAS № 5. Geography series, pp 5–12

Glazovskaya MA, Kasimov NS (1987) Landshaftno-geohimicheskie osnovyi fonovogo monitoringa prirodnoy sredyi (Landscape and geochemical bases of environment background monitoring). Reporter of MSU № 1. Geography series, pp 3–18

Grebenshchikov OS, Tishkov AA (1986) Geograficheskie zakonomernosti strukturyi i funktsionirovaniya ekosistem (Geographical regularities of geosystems structure and functioning). Nauka, Moscow, p 298

Iberla K (1980) Faktornyiy analiz (Factor analysis). Statistics, Moscow, p 398

Isachenko AG (1972) Puti sinteticheskogo izobrazheniya prirodnyih kompleksov, izmenennyih deyatelnostyu cheloveka (Ways of synthetic image on natural complexes changed by human activity) Synthesis problems in cartography. MSU, Moscow, pp 13–34

Isachenko AG (1974) O tak nazyivaemyih antropogennyih landshaftah (About so-called anthropogenous landscapes). Notion of RGS, pp 16–24

Isachenko AG (1981) Predstavlenie o geosisteme v sovremennoy fizicheskoy geografii (Idea of geosystem in modern physical geography). Notion of RGS 4(113):25–33

Isachenko AG (1990) Intensivnost funktsionirovaniya i produktivnost geosistem (Functioning intensity and efficiency of geosystems). Reporter of MSU № 5. Geography series, pp 17–23

Ioganson NK (1970) Klassifikatsiya antropogennyih landshaftov (Classification of anthropogenous landscapes). Reporter of LSU № 24, pp 14–25

Jeffers J (1981) Vvedenie v sistemnyiy analiz: primenenie v ekologii (Introduction to systems analysis: application in ecology). Mir, Moscow, p 213

Kalygin VP (2000) Promyishlennaya ekologiya (Industrial ecology). IIEPU, Moscow, 6–14 pp

Kasimov NS (1980) Geohimiya landshaftov zon razlomov (Geochemistry of landscapes on break zones). Vysshaya Shkola, Moscow, p 453

Kasimov NS (1990) Ekologo-geohimicheskaya otsenka gorodov (Ecological-geochemical assessment of the cities). Reporter of MSU № 3, Geography series, pp 22–27

Kasimov NS Perelman AI (1993) Geohimicheskie printsipyi ekologo-geohimicheskoy sistemyi gorodov (Geochemical principles of ecological-geochemical system of cities). Reporter of MSU № 3. Geography series, pp 16–21

Krauklis AA (1979) Problemyi eksperimentalnogo landshaftovedeniya (Problems of the experimental landscape studying). Nauka, Novosibirsk, p 120

Krauklis AA (1989) Vzaimodeystvie protsessov i struktur v geosistemah. Geografiya i prirodnyie resursyi (Interaction of processes and structures in geosystems. Geography and natural resources). Magazine geography and natural resources. 4:28–32

Makunina AA (1980) Funktsionirovanie i dinamika landshaftov (Functioning and dynamics of landscapes). Reporter of MSU № 5. Geography series, pp 2–14

Makunina TS (1983) K razvitiyu sovremennyih predstavleniy o geosistemah (To development of geosystem modern ideas). Reporter of MSU № 5. Geography series, pp 34–37

Mage D, Zali O (eds) (1992) Motor vehicle air pollution. Public health impact and control measures. Word Health Organization. Geneva, p 245

Metodicheskie rekomendatsii po geohimicheskoy otsenke istochnikov zagryazneniya okruzhayuschey sredyi (Methodical recommendations about geochemical assessment of environmental pollution sources) (1982) Vysshaya Shkola, Moscow, pp 74–82

Metodicheskie rekomendatsii po otsenke zagryazneniya territoriy gorodov himicheskimi elementami (Methodical recommendations about assessment of the chemical elements pollution in city territories) (1982) Vysshaya Shkola, Moscow, pp 11–19

Mickiewicz DF, Suslik YuYa (1981) Osnova landshaftno-geohimicheskogo rayonirovaniya (Basis of landscape and geochemical division). Kiev, p 238

Milkov FN (1967) Parageneticheskie landshaftnyie kompleksyi (Paragenetic landscape complexes). Scientific center of Voronezh department of GO USSR. Voronezh, pp 29–37

Milkov FN (1972) Voprosyi tipologii i kartografirovaniya antropogennyih landshaftov. Materialyi regionalnoy konferentsii «Antropogennyie landshaftyi tsentralnyih chernozemnyih oblastey i prilegayuschih territorii» (Typology and mapping issues of anthropogenous landscapes. Materials of the regional conference "Anthropogenous landscapes of the central black earth regions and adjoining territories"). Voronezh, pp 74–83

Milkov FN (1981) Basseyn reki kak paradinamicheskaya landshaftnaya sistema i voprosyi prirodopolzovaniya (River basin as paradynamic landscape system and questions of environmental management). Geography Nat Resour 4:247–252

Moiseenko TA (1989) Ekologo-geohimichekiy analiz promyishlennogo goroda (Ecological-geochemical analysis of the industrial city). Vysshaya Shkola, Moscow, pp 46–57

Nekhoroshev VP (1934) Kratkiy geologicheskiy ocherk territorii Bolshogo Altaya (Short geological sketch of the Big Altai territory). Kazakhstan, Alma-ata, p 420

Nikolaev VA (1978) Klassifikatsiya i melkomasshtabnoe kartografirovanie landshaftov (Classification and small-scale mapping of landscapes). MSU, Moscow, p 123

Nikolaev VA (1989) Landshaftnoe prostranstvo-vremya (metodologicheskie aspektyi) (Landscape space-time (methodological aspects)). Reporter of MSU № 2. Geography series, pp 24–27

Panin MS (2000) Himicheskaya ekologiya (Chemical ecology). Semipalatinsk state university named after Shakarim, Semipalatinsk, p 318

Perelman AI (1964) Geochemical landscapes (map of the USSR geochemical landscapes) (Geohimicheskie landshaftyi (karta geohimicheskih landshaftov SSSR)). Scale 1:20000000. Physiographic atlas of the world. Nauka, Moscow, p 64

Perelman AI (1966) Geohimiya landshafta (Landscape geochemistry). Vysshaya Shkola, Moscow, p 148

Perelman AI (1968) Geohimiya epigeneticheskih protsessov (Geochemistry of epigenetic processes (hyper genesis zone)). Subsoil, Moscow, pp 18–36, 43–54

Protasov VF (2000) Ekologiya, zdorove i ohrana okruzhayuschey sredyi v Rossii (Ecology, health and environmental protection in Russia. Finance and statistics). Moscow, pp 32–39

Reteyum AY (1975) Fiziko-geograficheskoe rayonirovanie i vyidelenie geosystem. Kolichestvennyie metodyi izucheniya prirodyi (Physiographic division and allocation of geosystems. Quantitative methods of nature studying). Mysl', Moscow, pp 5–27

Sochava VB (1978) Vvedenie v uchenie o geosistemah (Introduction to the geosystems doctrine). Nauka, Novosibirsk, p 318

Solntsev NA (1975) Formyi uporyadochennosti fiziko-geograficheskoy strukturyi (Orderliness forms of physiographic structure). In coll.: New in physical geography. Nauka, Moscow, pp 24-34

Solntsev VN (1981) Sistemnaya organizatsiya landshaftov: problemyi metodologii i teorii (System organizations of landscapes: problems of methodology and theory). Mysl', Moscow, p 215

Solntseva NA (1982) Geohimicheskaya ustoychivost prirodnyih geosistem k tehnogenezu. Dobyicha poleznyih iskopaemyih i geohimiya prirodnyih sistem (Geochemical resistance of natural geosystems to technogenesis. Mining and geochemistry natural system). Nauka, Moscow, pp 37–44

Shukputov AM (2001) Zagryaznenie: vidyi, problemyi i meryi. Ekologiya i ustoychivoe razvitie (Pollution: types, problems and measures. Ecology and sustainable development). 8:2–5

Svirezhov YM (1982) Matematicheskie modeli v ekologii. Chislo i myisl (Mathematical models in ecology. Number and thought). № 5, vol 5. pp 16–55

Vladimirov VV, Mikulina EM, Yargin ZN (1986) Gorod i landshaft. Problemyi, konstruktivnyie zadachi i resheniya (City and landscape. Problems, constructive tasks and decisions). Mysl', Moscow, p 336

Volkova VG, Davydov ND (1987) Tehnogennaya transformatsiya landshaftov (Technogenic transformation of landscapes). Nauka, Novosibirsk, pp 54–69

Vorobeychek EA (1994) Ekologicheskoe normirovanie tehnogennyih zagryazneniy v nazemnyih ekosistemah (Ecological rationing of technogenic pollution in land geosystems). Vysshaya Shkola, Moscow, pp 55–70

Tehnogennyie potoki veschestva v landshaftah i sostoyanie ekosistem) Technogenic streams of substance in landscapes and condition of geosystems) (1981) MSU, Moscow, p 351

The program of development of the East Kazakhstan Region territory on 2016–2020

Weyant YP (1997) Insights from integrated assessment. IPIECA Symposium on the critical issues in the economics of climate change. London, pp 245–276

Chapter 3
Division of the Territory of East Kazakhstan According to the Level of Anthropogenic Impact

3.1 Methods in Landscape and Ecological Division Based on Criterion Function

We used methodical developments on landscape ecological zoning, on the basis of target function from staff work of department of environment geoecology and monitoring [Pavlichenko L.M., Urikbayeva Z.S., Chigarkin A.V. An expert assessment of anthropogenous impact on environment of the Kyzylorda region by means of target function//the KazNU Journal. Series on ecology. 2002. № 1 (10). 70–74 p.]. We will introduce its short summary, which reflects our research.

Urgent problem of geoecology (landscape ecology) consists in spatial regularities studying of interaction between human society and nature. The purpose of landscape ecology as scientific discipline is studying ecological conditions of the environment concerning people's activity in geosystem borders, natural and natural-economic regions of various taxonomical advantages (Gerasimov 1986).

In the methodical guide on carrying out ecological zoning of the Republic of Kazakhstan territory /96/ on the basis of the Cabinet Council Provision of the RK № 548 from July 29, 1993 year "On urgent measures to streamline the ecological zoning of the Republic of Kazakhstan" it is noted, that ecological zoning has to be directed to regions delimitation of the republic with adverse environment to development of governmental activities on restoration of protective and rehabilitation measures for the population living there (Chizhov et al. 1995; Methodical guide to carrying out ecological division of the Republic of Kazakhstan territory 1995).

Thus, the tasks set for ecological zoning define the main orientation on allocation of crisis regions in ecological aspect, for purpose of scientific justification of legal, economic and administrative decisions, directed to stabilization and restoration of people's health state and environment.

© Springer Nature Singapore Pte Ltd. 2018
S. Bayandinova et al., *Man-Made Ecology of East Kazakhstan*,
Environmental Science and Engineering, https://doi.org/10.1007/978-981-10-6346-6_3

Ecological situation in natural and anthropogenous systems depends on many factors: environment and resistance degree of technogenesis geosystems, character and intensity of anthropogenous impact on environment, reaction of the population to anthropogenesis manifestation and so forth. Listed factors, defining features of geosystems as vital circle of people, are unevenly distributed on space of geographical envelope. In formation of ecological situations concerning natural and anthropogenous systems, there is a big and various roles of technogenesis, that causes the necessity of knowledge about regional features, technogenic impact on its environment and management. Very close connection is found concerning anthropogenesis manifestation with specifics of local environment of physiographic regions.

For management of technogenic processes, with the aim of greening production purpose, it is also necessary to know sources of technogenesis, features of their placement, nature of impact on the environment, manifestation conditions of this influence and its distribution in space. The last one is important, especially for various aspects of the territorial organization concerning productive forces. In this regard, there is a need of environmental problems research in spatially—territorial aspect, i.e. on the basis of geoecological zoning. In this case, such basic provisions of geographical approach are fully implemented to environmental problems study as complexity (joint studying of natural and technical systems) and territoriality (spatial specifics).

Geoecological zoning serves for justification of activities on protection and rational use of the environment and assumes the solution of following private tasks:

1. Studying of natural system features, detection of regularities on their spatial and structural variability.
2. Studying of geosystems reaction to natural and technogenic influences, and assessment of environmental and engineering risk.
3. Identification of technogenic influence sources, their ecological characteristic, determining the extent and nature of the changes in the natural environment caused by them.
4. Assessment of the natural systems' current state, extents of its change under the technogenesis influence, identification of problem ecological situations and tendencies of an ecological situation development.

Due to the need of these tasks solution, there is a demand for the whole set of maps, which target loading and informational content changes according to problems of concrete stage, therefore the full list of maps can vary, however I.A. Avessalomova suggests to allocate several interconnected blocks: inventory maps, estimated, expected landscape and ecological, and maps of the recommended nature protection and other activities, moreover assessed maps play an important role with displayed results of private and integrated assessments on current ecological state of landscapes. These maps serve as justification to compile various applied maps.

Assessment of natural systems condition is carried out on the basis of comparison with reference systems, and it is possible at measurement of extent on quantitative value excesses concerning ecological factors. Various standards are often acted as standards of optimum environment condition.

While assessing the impact of environmental factors on the state of the natural environment, the interrelations of the influencing factors with the response of geoecosystems should be taken into account. There is a problem of threshold value fixing on reaction of biotic component, when influence of ecological factor becomes negative.

As for each live organism there is an ecological niche of environment optimum conditions, the fact of values optimum range existence for each influencing factor is quite natural, at the same time negative ones will be deviations both towards increase, and towards decrease from optimum conditions.

For this reason, one of the key moments at impact assessment is development of private assessment scales, in the course of drawing up is necessary to bring two groups of indicators, one of which characterizes condition of ecological factor, another—a condition of geoecosystems biotic component. The logic of assessment is given in /98/and looks as follows:

Studying of factor and its features → collection of information on ecological communications and identification of responses on biotic components → the choice of criteria and development of private assessed scale → impact assessment of factor and establishment of spatial variation, which can be taken out on separate maps.

So, the problem of impact assessment on concrete ecological factor comes down to the choice of parameters set, most objectively characterizing this factor and creation of rating scale.

At complex assessment, it is necessary to consider not only intensity of each ecological factor influence, but also its role in formation of favorable or negative living condition of biosystems. The simplest opportunity of such complex assessment gives linear target function (Krauklis 1986).

$$\Sigma infl. = a_1 \times f_1 + a_2 \times f_2 + \ldots + a_n \times f_n,$$

here f_i—value of concrete ecological factor in observation point, a_i—the weight coefficient considering an orientation (plus or minus) and the importance (weight) of this factor in formation of total influence level.

As in the given equation, effects on interaction of influence factors are not considered, the assessment accuracy will go up at increase in number of the recorded factors. Methods of engineering ecology (in particular, at creation of the main ecological table) show, that satisfactory on accuracy the number of factors equal to five.

While implementing impact assessments, there is a fact to face the lack quantitative data on the influencing ecological factors, in this case it is resorted to the methods of expert assessment generalizing the accumulated experience on research of influence considering various ecological factors. However, even in this case there are still problems of the most significant factors choice and development of private and integrated scales.

3.2 Map of Sources of Anthropogenic Impact on Geosystems in East Kazakhstan

As it was mentioned in justification of division technique on anthropogenous loading, results of the first information stage of ecological assessment can be expressed in inventory maps, in which objects of influence and ecological factors are reflected. Example of such map is "The schematic map of anthropogenous loading" as it is taken out by the corresponding signs of manifestation of anthropogenous impact with different types. These manifestations can carry local (concentrated) or area character, at the same time the task of extent assessment of anthropogenous factors influence is not set on geosystems. To the first type of influences, it is possible to carry agricultural grounds, to the second—mining enterprises extracting minerals in the opened and closed way, concentrating factories, tailing dams.

At map development, they used already available maps, reflecting the nature of land use on territories, distribution of landscapes, mineral deposits, and the actual material collected in various ecological, hydrogeological and statistical services of East Kazakhstan. Map published in 1998 "The East Kazakhstan region M1:1,000,000" is chosen as basis of map construction, where administrative borders, settlements, hydrography are reflected. Besides, they used published maps such as "Map of land use of the East Kazakhstan and Semipalatinsk regions M1:1,000,000"; "The schematic map of an enterprises arrangement on mineral and raw complex—environment pollutants", developed by the East Kazakhstan Altai department of Geology Institute named after K.I. Satpayev and the East Kazakhstan regional territorial administration of environmental protection (EKRTDEP), T.S. Kertishov's map and etc. "Especially protected natural territories of the Republic of Kazakhstan M 1:1,000,000". Furthermore, materials of Geology Institute of AS RK, Botanical Institute of AS RK, Soil science Institute of AS RK, RSE "Kazgidromet" were used.

Parameters chosen from these maps, reflecting anthropogenous load of the territory were applied on basis with an imposing method. As a result of points combination and the manifestation areas of anthropogenous loading factors "The map of sources on anthropogenous impact of East Kazakhstan geosystems" was constructed.

Map symbols to the figure—(1) Contours with flowers allocate agriculture factors: arable lands, irrigated arable lands, solonetzic arable lands, upland pastures, flooded pastures, pastures littered with stones, pastures littered with stones and grown bushes, pastures with grown bushes, pastures overgrown with bush solonetzic, forested pastures, solonetzic pastures, salty pastures, flooded haymakings, upland haymakings, tape pine forests on sand, wood, swamp, saline soils. (2) Allocated with badges: the developed fields and in the process of development (3) Industrial hubs are allocated by areas covered with shading and designated by the Arab figures: *(1) Shemonaykhinsky; (2) Ridder; (3) near the Ertis; (4) Serebryansko-Belogorsky; (5) Zyryanovsky.* (4) Especially protected natural territories are allocated by contours and the Roman figures: I National nature wildlife park—Semey ormany; national nature parks: II West Altai, III Markakol, state natural wildlife areas: IV Nizhniy—Turgusunsky, V Kulundzhunsky, VI Tarbagatai; VII State natural park Katon-Karagai. The Semipalatinsk nuclear test site is allocated by red dotted line (Konstantinov and Chelidze 2001).

Thus, anthropogenous loading on the map is presented by manifestations on the main sources of anthropogenous influence: agriculture—arable lands, haymakings, pastures and industrial hubs—mines, mineries, concentrating factories, iron and steel works, tailing dams, zone of the Semipalatinsk nuclear test site. Also on the map manifestations of natural factors are taken out, promoting environment self-restoration, such as tape pine forests, woods, swamps, saline soils, water objects. Besides, contours of SPA (specially protected area) of different level are allocated. Apparently from the Fig. 2.8, we can see that the map of sources of anthropogenous influence, the main anthropogenous loading is concentrated in zone of the densest accommodation, therefore as a separate factor of anthropogenous pressure population density was not considered.

3.3   Division of the Territory of East Kazakhstan According to the Anthropogenic Impact

The successful management of region economy now demands definition of not only the quantitative part of natural resources, but also their ecological state, as quality of the environment becomes the limiter of economy development. For East Kazakhstan quality of the population life is defined first of all by state, natural geosystems and water resources, on which development of recreational zones and tourist routes is possible, and also defining development of agricultural branches, mining and processing industry, metallurgical combines. Powerful psychological factor is existence of Semipalatinsk nuclear test site. Thus, anthropogenous impact on geosystems has multidirectional character and that problem of complex assessment influence is sharply allocated.

For target function, the following indicators were considered:

- Arable lands,
- Irrigated arable lands,
- Solonetzic arable lands,
- Upland pastures,
- Flooded pastures,
- Pastures littered with stones,
- Pastures littered with stones and grown bushes,
- Pastures with grown bushes,
- Pastures overgrown with bush solonetzic,
- Forested pastures,
- Solonetzic pastures,

- Salty pastures,
- Flooded haymakings,
- Upland haymakings,
- Tape pine forests on sands,
- Wood,
- Swamp,
- Saline soils,
- Areas of open water surface,
- Open-end enterprises,
- Enterprises of underground mining,
- Forecasted enterprises for explored deposits,
- Processing enterprises,
- Ore mining and processing factories.

As the majority of anthropogenous loading characteristics have qualitative feature, which are submitted on inventory map, and for creation of complex assessment (results of target function) we used system of parameters coding on the basis of net model of the East Kazakhstan territory M 1:1000000. In order to create the net model, Schematic map was divide into 187 blocks by regular grid, an each block represents a cell quarter of standard coordinate network M1:1000000 $(1° \times 1°)$.

All parameters used for complex assessment are divided by us into areas: swamps, saline soils, areas of open water surface, arable land, irrigated arable land, solonetzic arable land, upland pastures, flooded pastures, pastures littered with stones, pastures littered with stones and grown bushes, pastures with grown bushes, pastures overgrown with bush solonetzic, salty pastures, solonetzic pastures, flooded haymakings, upland haymakings, tape pine forests on sands, forested pastures, woods; simple local ones: ore mining and processing factories (OMPE), enterprises of manufacture, and locally classified ones: enterprises of open-cast mining, enterprises of underground mining, expected enterprises for the reconnoitered fields.

As a unit to measure simple area characteristics, we used the conventional unit equal to $\frac{1}{64}$ part of the block. Point assessment of such characteristics in each block was defined by conditional area of their distribution in the block.

For simple local characteristics, the number of their manifestations in the block was counted and multiplied by role point of each manifestation, taking into account influence of enterprise technological scheme on environment (6 points for OMPE, from 7 to 10 points for enterprises of reprocessability).

For local classified signs, points paid off on classification point (to enterprises with open form of minerals development, points ranged from 8 to 10, and with underground mining—from 5 to 7 was appropriated) and 10 for areas of the developed fields.

For area characteristics, the area of their manifestations in the block as a share from total area of the block was counted, and it was multiplied by role point of each manifestation taking into account influence of each sign on the environment

(Irrigated arable lands—10; Arable lands—9; Solonetzic arable lands—8.5; Forested pastures—8; Upland pastures—7; Pastures with grown bushes—6.5; Solonetzic pastures—6; Flooded pastures—5.5; Pastures littered with stones—; Pastures littered with stones and grown bushes—4.5; Pastures overgrown with bush solonetzic—4; Salty pastures—3.5; Flooded haymakings—3; Upland haymakings—2.5; Saline soils—2. For area signs promoting restorable landscapes, values of points were appropriated according to their role in self-restoration: Tape pine forests on sands—8; Woods—10; Areas of open water surface—10.

Matrix of values on technogenic influence factors for geosystems of East Kazakhstan, representing net model of basic data on each block is given in Table 3.1.

For creation of target function, it is necessary to assess a role of each factor (through weight coefficient) and an orientation of its action. Impact orientation on the environment of each factor in target function was considered by a sign: factors, interfering negative change of geoecological situation include target function with a minus sign.

Currently pastures can be such factors in our territory, which partially restore the clearing role through significant increase in green material in the period of pasture lackness. Therefore in target function this factor has to have negative loading (−0.5). Haymakings are much less, than pastures, and exposed to anthropogenous pressure, that's why their role in self-restoration of landscapes is higher and their load is accepted by −0.7. Traditionally clearing role belongs to the forests, presented in our case by tape pine forests on sands (−0.8), to the woods (−1.0) and open water surface (−1.0). These factors also have high negative loading in target function [192–195].

Impact of the irrigated massifs on environment state is ambiguous. As positive influence it is possible to accept air moistening, existence of the green material promoting air clarification, completion of ground water reserves. Negative consequences are violation of soil cover, secondary salinization, flooding, pollution of ground waters surplus of pesticides and fertilizers. For the accounting of irrigated massifs complex influence in target function, contribution of this factor was assessed at total technogenic impact on landscape and ecological situation of the region by coefficient 0.5.

Table 3.1 Classification of target function values for assessment of technogenic impact on landscape and ecological situation

№ of classes	Characteristics of classes	Limits of target function value (on points)
A	Practically there is no anthropogenous influence	<10
B	Insignificant anthropogenous influence	−10÷+5
C	Noticeable anthropogenous influence	+5÷+20
D	Strong anthropogenous influence	+20÷+35
E	Zones of ecological crisis	>+35

A special factor is designated as the Semipalatinsk nuclear test site. Its territory is not allocated through the parameters including target function: there is no considerable industrial and agrarian construction pollution, and river pollution is not allocated, natural radioactive pollution is not noted. Thus, a traditional set of ecological characteristics of environmental pollution does not reflect influence of the ground for which accounting additional factors are necessary. Such factors are violation of firm based landscapes, soils pollution, and considerable psychological tension of population. For the accounting of this factor, high positive weight loading is also accepted (0.7).

Now it is possible to write down a type of target function for complex assessment of anthropogenous impact on landscape and ecological situation of East Kazakhstan (AA infl.):

AA infl. = 0.5 arable lands +0.5 irrigated arable lands +0.5 solonetzic arable lands −0.5 upland pastures −0.5 flooded pastures −0.5 pastures littered with stones −0.5 pastures littered with stones and grown bushes −0.5 pastures with grown bushes −0.5 pastures overgrown with bush solonetzic −0.5 forested pastures −0.5 solonetzic pastures −0.5 salty pastures −0.7 flooded haymakings −0.7 upland haymakings −0.8 tape pine forests on sand −1 wood −0.7 swamp +0.3 saline soils −1 areas of open water surface +1 enterprises of open-cast mining +0.5 enterprises of underground mining +1 expected enterprises for the reconnoitered fields +1 enterprises of manufacture +1 OMPE + 0.7 SNTS.

Calculation results of target function for each block are presented in the 26th column of Table 3.2.

The following evaluation stage is classification of target function values according to limits of each factor changes, which will allow to get rid of measure units' influence and will give the chance of comparison with earlier received results of assessment.

Maximum value of complex assessment in the block will be defined by the maximum value of simple area factor on condition of full filling the area of the block (10 points), plus the sum of negative local factors influence, which total impact is defined from condition that the maximum influence of each local factor cannot exceed influence of half area (5 points for each). In this case the maximum number of points will make 55. Minimum complex assessment will turn out at the maximum influence of factors promoting stabilization of geoecosystems. By analogy with the maximum assessment, we will receive value—20 (as only two local factors of situation stabilization are considered).

Borders of classes can be accepted in compliance with methods of engineering ecology according to the following Table 3.1.

According to this table, following target function values have to be borders of classes:—10 (between the 1 and 2 classes), 5 (between 2 and 3 classes), 20 (between 3 and 4 classes), 35 (between 4 and 5 classes). Received target function values were attributed to the block centers, and after creation of isolines, to the corresponding borders of classes, the scheme of division on intensity of techno-genic impact on landscape and ecological situation of East Kazakhstan (Fig. 3.1) was received.

Table 3.2 Results data on target function of the East Kazakhstan territory

№ of blocks	Arable lands	Irrigated arable lands	Solonetzic arable lands	Upland pastures	Flooded pastures	Pastures littered with stones	Pastures littered with stones and grown bushes	Pastures with grown bushes	Pastures overgrown with bush solonetzic	Forested pastures	Solonetzic pastures	Salty pastures	Flooded haymakings
1	2	3	4	5	6	7	8	9	10	11	12	13	14
1-1													
1-2	4.8												
1-3				1.3				0.8					
1-4													
1-5													
1-6													
1-7													
1-8													
1-9													
1-10													
1-11													
1-12													
1-13													
1-14													
1-15													
1-16													
1-17													
2-1													
2-2	1.4			2.8									
2-3	2.2			2.8						0.8		0.4	
2-4	5.9			2									
2-5	5.7							2.2					

(continued)

Table 3.2 (continued)

№ of blocks	Arable lands	Irrigated arable lands	Solonetzic arable lands	Upland pastures	Flooded pastures	Pastures littered with stones	Pastures littered with stones and grown bushes	Pastures with grown bushes	Pastures overgrown with bush solonetzic	Forested pastures	Solonetzic pastures	Salty pastures	Flooded haymakings
2–6	2.4							4.6					
2–7													
2–8													
2–9													
2–10													
2–11													
2–12													
2–13													
2–14													
2–15													
2–16													
2–17													
3–1	1.3			3.5				0.5					0.2
3–2	2.7			3.4									
3–3	3.3			1.9									
3–4	3.6												
3–5	7.7			0.6									
3–6	3.8			1.2				2.1					
3–7	7.1			1.2									
3–8													
3–9													
3–10				2.5									
3–11				5.3									

(continued)

Table 3.2 (continued)

№ of blocks	Arable lands	Irrigated arable lands	Solonetzic arable lands	Upland pastures	Flooded pastures	Pastures littered with stones	Pastures littered with stones and grown bushes	Pastures with grown bushes	Pastures overgrown with bush solonetzic	Forested pastures	Solonetzic pastures	Salty pastures	Flooded haymakings
3-12													
3-13													
3-14													
3-15													
3-16													
3-17													
4-1	1.3			1.4									
4-2	0.8			2.3									
4-3	0.7			2.6				0.4					
4-4	5.1			0.5				0.3					
4-5	5.1			0.8									
4-6	5.6			2									
4-7	4.4			2.9									
4-8				0.9									
4-9				0.1									
4-10				0.9									
4-11				2.8									
4-12													
4-13													
4-14													
4-15													
4-16													
4-17													

(continued)

Table 3.2 (continued)

№ of blocks	Arable lands	Irrigated arable lands	Solonetzic arable lands	Upland pastures	Flooded pastures	Pastures littered with stones	Pastures littered with stones and grown bushes	Pastures with grown bushes	Pastures overgrown with bush solonetzic	Forested pastures	Solonetzic pastures	Salty pastures	Flooded haymakings
5-1				5.9									
5-2	1.4			5.3									
5-3	1.7			2.7									
5-4	0.7	0.6		1.5				0.2	0.1	0.5		0.6	
5-5	0.8			1.4					0.07		0.6	0.7	
5-6	2.8	0.4		1.5			0.5						
5-7	4.4			1.6			0.6						
5-8	5.6			0.4									
5-9	4.2			0.2				0.4					
5-10	0.8			2.9									
5-11				2									
5-12				5.3									
5-13													
5-14													
5-15													
5-16													
5-17													
6-1	1.1			3.2		5.2					0.2		
6-2	0.9			4.4		0.8					1.2		
6-3	1.5			5.4		1.2					0.2		
6-4				0.5					0.3				
6-5	0.9			4.6		1.9	0.2	1.3			0.9		
6-6	1.5					0.9							

(continued)

Table 3.2 (continued)

№ of blocks	Arable lands	Irrigated arable lands	Solonetzic arable lands	Upland pastures	Flooded pastures	Pastures littered with stones	Pastures littered with stones and grown bushes	Pastures with grown bushes	Pastures overgrown with bush solonetzic	Forested pastures	Solonetzic pastures	Salty pastures	Flooded haymakings
6-7	3.8			1.2		0.9	1.1						
6-8	1.9			5.1									
6-9	2.8			2.7				0.4					
6-10	0.7			1.2									
6-11				4.4									
6-12	0.5			0.7									
6-13				0.8									
6-14													
6-15													
6-16													
6-17													
7-1	0.1			0.9		4.7						0.9	
7-2	0.7			0.7		3.3						1.3	
7-3			1			3.4						0.6	
7-4	0.5			0.1		4							
7-5	0.7			0.3		4.5	0.2						
7-6	0.8			0.7		2.9		1.4					
7-7	0.8			0.2		2.5		2.2					
7-8	0.5					1.1		4.8					
7-9	0.5			5.9				0.2	0.1				
7-10				3.9									
7-11				4.7									
7-12	1.6			5.7									
7-13	0.5			2.4									

(continued)

Table 3.2 (continued)

№ of blocks	Arable lands	Irrigated arable lands	Solonetzic arable lands	Upland pastures	Flooded pastures	Pastures littered with stones	Pastures littered with stones and grown bushes	Pastures with grown bushes	Pastures overgrown with bush solonetzic	Forested pastures	Solonetzic pastures	Salty pastures	Flooded haymakings
7-14				2.1						2.1			
7-15				5						2.3			
7-16				3.6						2.1			
7-17										2.1			
8-1	0.1			2.4		3.5							
8-2	0.4			3.1		2.8							
8-3			0	0.5		3.5	0.5					0.5	
8-4	0.4		0	1.4		2.8					0.8		
8-5	0.1			0.2		5.2							
8-6	0.5				5.2								
8-7	1.3			0.8		3.4							
8-8	0.5			5.1		0.9							
8-9				7									
8-10	0.1			6									
8-11	2.3			2.7									
8-12	1.3			4.3									
8-13	9.2	2		3.4									
8-14	1.1			4.8									
8-15	0.7			4.9									
8-16				4.9									
8-17				6.1				0.4					
9-1	1.1			0.4		3.9							
9-2	0.3			6.3		0.3							

(continued)

Table 3.2 (continued)

№ of blocks	Arable lands	Irrigated arable lands	Solonetzic arable lands	Upland pastures	Flooded pastures	Pastures littered with stones	Pastures littered with stones and grown bushes	Pastures with grown bushes	Pastures overgrown with bush solonetzic	Forested pastures	Solonetzic pastures	Salty pastures	Flooded haymakings
9–3	0.7			4.7		1.4							
9–4	0.7			0.7		3.9				0.5			
9–5	0.8			3.8		1.9							
9–6				2.7		3.4							
9–7				5.3		1.4							
9–8	0.4			5.3		1.1							
9–9	3.7			2.7		0.5							0.2
9–10	4.5			3.1									
9–11	0.9			3.9						0.6			
9–12				7									
9–13				6.1									
9–14				5.9						2.1			
9–15				3.4						1.8			
9–16	0.4			3.7		0.6							
9–17													
10–1	1.3					4.6				0.1			
10–2	0.7					4.9							
10–3	0.1			5.3		0.5			0,6				
10–4	0.8			1.4		3.3							
10–5	0.9			0.4		4.1					0.4		
10–6	0.4			50		1.3					0.4		
10–7	0.4			5.3		1.1							
10–8	1.3			0.4		4.3							

(continued)

Table 3.2 (continued)

№ of blocks	Arable lands	Irrigated arable lands	Solonetzic arable lands	Upland pastures	Flooded pastures	Pastures littered with stones	Pastures littered with stones and grown bushes	Pastures with grown bushes	Pastures overgrown with bush solonetzic	Forested pastures	Solonetzic pastures	Salty pastures	Flooded haymakings
10–9	1.6			3.6	0.5	0.8							0.3
10–10	2.3			3.7	0.7								0.2
10–11	0.9	1.9		2.6		1							
10–12	0.5			4.7		1.3							
10–13				7									
10–14				6.8									
10–15				5.8									
10–16				5.6									
10–17													
11–1	0.7					0.9	3.7					0.1	
11–2	0.4					4.1	1						
11–3	4.9			2.3		0.5	0.5						
11–4	1.1			5		0.5						0.2	
11–5	0.5			1.3		3.9						0.2	
11–6	0.5			0.9		4.4							
11–7	0.4			1.2		4.3							
11–8				1.4	5.2	0.6							
11–9	1.5			4.9				0.8					
11–10				3.2				0.7					
11–11				3.7									
11–12				5.3									

(continued)

Table 3.2 (continued)

№ of blocks	Arable lands	Irrigated arable lands	Solonetzic arable lands	Upland pastures	Flooded pastures	Pastures littered with stones	Pastures littered with stones and grown bushes	Pastures with grown bushes	Pastures overgrown with bush solonetzic	Forested pastures	Solonetzic pastures	Salty pastures	Flooded haymakings
11-13				6.2									
11-14				5.6									0.6
11-15				0.3									
11-16													
11-17													

№ of blocks	Upland haymakings	Tape pine foress on sand	Wood	SNTS	Saline soils	Areas of open water surface	Enterprises of open-cast mining	Enterprises of underground mining	Expected enterprises for the reconnoitered fields	Enterprises of manufacture	Mining and processing factories	EAAI (Ecological assessment of anthropogenuous influence)
1	15	16	17	18	19	20	21	22	23	24	25	26
1-1												0
1-2												0
1-3		1.3										-0.65
1-4												0
1-5												0
1-6												0
1-7												0
1-8												0
1-9												0
1-10												0
1-11												0
1-12												0

(continued)

Table 3.2 (continued)

№ of blocks	Upland haymakings	Tape pine forests on sand	Wood	SNTS	Saline soils	Areas of open water surface	Enterprises of open-cast mining	Enterprises of underground mining	Expected enterprises for the reconnoitered fields	Enterprises of manufacture	Mining and processing factories	EAAI (Ecological assessment of anthropogenuous influence)
1-13												0
1-14												0
1-15												0
1-16												0
1-17												0
2-1												0
2-2		4.3										-4.42
2-3												-1.34
2-4		0.1										0.69
2-5												0.61
2-6												-1.58
2-7												0
2-8												0
2-9												0
2-10												0
2-11												0
2-12												0
2-13												0
2-14												0
2-15												0
2-16												0
2-17												0
3-1						1.7						-3.45

(continued)

Table 3.2 (continued)

№ of blocks	Upland haymakings	Tape pine forests on sand	Wood	SNTS	Saline soils	Areas of open water surface	Enterprises of open-cast mining	Enterprises of underground mining	Expected enterprises for the reconnoitered fields	Enterprises of manufacture	Mining and processing factories	EAAI (Ecological assessment of anthropogenuous influence)
3-2		2						1				-1.84
3-3			3.4									-3.36
3-4			5.8									-4.72
3-5							4					5.43
3-6			0.5									-1.01
3-7								4				3.58
3-8												0
3-9			10									-10
3-10			6.5						2			-5.95
3-11			2.4									-5.05
3-12												0
3-13												0
3-14												0
3-15												0
3-16												0
3-17												0
4-1												0
4-2			4.7			1.7		5				-4.36
4-3		0.2	4.4			0.8						-6.47
4-4			4.8			0.3						-6.34
4-5			3.3									-2.02
4-6			2.5			0.4						-1.77
4-7						0.5	3			7	6	14.7

(continued)

Table 3.2 (continued)

№ of blocks	Upland haymakings	Tape pine forests on sand	Wood	SNTS	Saline soils	Areas of open water surface	Enterprises of open-cast mining	Enterprises of underground mining	Expected enterprises for the reconnoitered fields	Enterprises of manufacture	Mining and processing factories	EAAI (Ecological assessment of anthropogenuous influence)
4–8			0.2			0.6						−0.93
4–9			8.6					4	10			1.85
4–10			9.8					2				−8.8
4–11			8.7					1	5			−4.12
4–12			6									−7.4
4–13												0
4–14												0
4–15												0
4–16												0
4–17												0
5–1												0
5–2												−2.53
5–3						0.5	2					−1.11
5–4	0.09					1.3	3					−0.72
5–5		2.9				1.6	2					−3.77
5–6		2.9				0.6						−2.91
5–7	0.2					0.6	4	2				3.94
5–8						1.2	2	4	1			2.76
5–9			3.8									−1.59
5–10			7.8			0.9		3	10	13	12	24.6
5–11			5.8					11	1			−0.62
5–12			7.1					6				−5
5–13			2.5									−5.15

(continued)

Table 3.2 (continued)

№ of blocks	Upland haymakings	Tape pine foress on sand	Wood	SNTS	Saline soils	Areas of open water surface	Enterprises of open-cast mining	Enterprises of underground mining	Expected enterprises for the reconnoitered fields	Enterprises of manufacture	Mining and processing factories	EAAI (Ecological assessment of anthropogenuous influence)
5-14												0
5-15												0
5-16												0
5-17												0
6-1							1	1				-1.36
6-2							6	1				3.26
6-3							3					-0.2
6-4							3					0.03
6-5							8					5.16
6-6						0.2	7					4.21
6-7							12					10.3
6-8						0.5						-2.48
6-9			0.9			1.4	4	2		41	12	48.9
6-10			7.5					2				-6.89
6-11			3.8									-6
6-12			8.4									-8.6
6-13			8.8				2	6				-4.52
6-14												0
6-15												0
6-16												0
6-17												0
7-1												-2.77
7-2												-2.24

(continued)

Table 3.2 (continued)

№ of blocks	Upland haymakings	Tape pine forests on sand	Wood	SNTS	Saline soils	Areas of open water surface	Enterprises of open-cast mining	Enterprises of underground mining	Expected enterprises for the reconnoitered fields	Enterprises of manufacture	Mining and processing factories	EAAI (Ecological assessment of anthropogenuous influence)
7-3								1				−1.6
7-4						0.5		3				−1.1
7-5						0.2	3					−0.15
7-6						0.5	1	5				0.91
7-7						0.8	14					9.86
7-8			0.4				3	3				1.68
7-9							8					4.03
7-10			4.1			0.4					12	4.35
7-11			0.8			2.5						−5.65
7-12								3		13	6	16.3
7-13	0.09		5.5					8				−2.41
7-14			4.4					1				−5.95
7-15												−3.65
7-16			2.2									−5.05
7-17												−1.05
8-1												−2.92
8-2								2				−2.03
8-3							3					−0.07
8-4												−2.26
8-5												−2.67
8-6												−2.45
8-7						1.3	1					−1.88
8-8						0.3						−3.15

(continued)

Table 3.2 (continued)

№ of blocks	Upland haymakings	Tape pine forests on sand	Wood	SNTS	Saline soils	Areas of open water surface	Enterprises of open-cast mining	Enterprises of underground mining	Expected enterprises for the reconnoitered fields	Enterprises of manufacture	Mining and processing factories	EAAI (Ecological assessment of anthropogenuous influence)
8-9							4					-3.5
8-10			1.3									-0.31
8-11			2.9			0.5						-4.06
8-12			0.5			1.9						-4.16
8-13	0.1		1.4									0.19
8-14			1.9					1				-3.42
8-15			2.2									-4.44
8-16			3									-5.45
8-17			1.3									-4.35
9-1							3					0.68
9-2												-3.21
9-3							1					-1.85
9-4							1					-1.35
9-5							2					-2.61
9-6							10					-0.89
9-7												6.01
9-8								8				0.92
9-9								4				1.12
9-10						0.5		6				2.3
9-11						1.9	3		1			-0.17
9-12								6				-0.75
9-13			1.3									-4.35
9-14			0.9			0.6		2				-3.3

(continued)

Table 3.2 (continued)

№ of blocks	Upland haymakings	Tape pine forests on sand	Wood	SNTS	Saline soils	Areas of open water surface	Enterprises of open-cast mining	Enterprises of underground mining	Expected enterprises for the reconnoitered fields	Enterprises of manufacture	Mining and processing factories	EAAI (Ecological assessment of anthropogenuous influence)
9-15			0.8			1.7						-5.25
9-16			0.9									-3.83
9-17							3					2.43
10-1												-1.91
10-2												-2.29
10-3												-3.17
10-4												-2.31
10-5												-2.18
10-6							2					-23.6
10-7												-3.08
10-8												-1.96
10-9							1					-1.01
10-10						0.3	3					0.39
10-11						1.4						-2.36
10-12						0.3	1	3				-0.46
10-13												-3.5
10-14			0.2			0.2	6					1.87
10-15			1.8				6					1.06
10-16			2									-4.8
10-17								1				0
11-1												-1.59
11-2												-2.43
11-3												0.07

(continued)

Table 3.2 (continued)

№ of blocks	Upland haymakings	Tape pine forests on sand	Wood	SNTS	Saline soils	Areas of open water surface	Enterprises of open-cast mining	Enterprises of underground mining	Expected enterprises for the reconnoitered fields	Enterprises of manufacture	Mining and processing factories	EAAI (Ecological assessment of anthropogenuous influence)
11-4												-2.52
11-5							1					-1.56
11-6												-2.5
11-7							2					-0.74
11-8							2					-2.16
11-9												-2.4
11-10	0.09				0.5	2						-3.86
11-11						4.7						-6.55
11-12					0.3	1.4						-3.96
11-13					0.1	4.7						-7.77
11-14												-3.22
11-15												-0.15
11-16												0
11-17												0

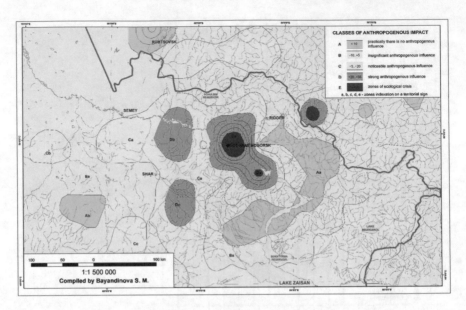

Fig. 3.1 Map zoning of anthropogenous impact on the East Kazakhstan territory

Apparently from the Fig. 2.9, distribution of target function reflects allocation across the territory of anthropogenous loading. Practically all territory is noted by class zone of B and C of the technogenic pollution characterizing by average level of insignificant and noticeable anthropogenous influence.

Class zone B of insignificant anthropogenous impact covers practically all territory western, edge of northern, east and central part of East Kazakhstan, which is named on index zones of territorial sign with binding to natural objects and settlements: Semey-Ridder-Shar. This zone includes specially protected natural sites of the national nature wildlife park—Semey ormany, located in West Altai.

Zone of noticeable anthropogenous impact covers three centers of distribution, which is named on index zones of territorial sign: Uba-Kokpekty-Kurchum, Karabulak, Karaganayryk.

Zone of the maximum anthropogenous impact of E class—ecological crisis covers the nucleus territory of Priertisky thoroughly soaped knot, which is named on territorial sign as Ust-Kamenogorsk, the main factor of anthropogenous loading, first of all processing industry: lead-zinc plant, Ust-Kamenogorsk iron and steel works, titanium-magnesium plant, and etc. (block 9–6). Zone of the maximum anthropogenous impact is covered by the zone of strong anthropogenous impact having three centers: Ulba-Taintin (blocks 5–8, 5–9, 5–10, 6–8, 6–10, 7–8, 7–9, 7–11) In the predominance of the industry factor, Shchulbin (blocks 5–6, 5–7, 6–6, 6–7), Zharly-Shar (blocks 8–6, 8–7) experiencing agricultural strain of arable lands and pastures.

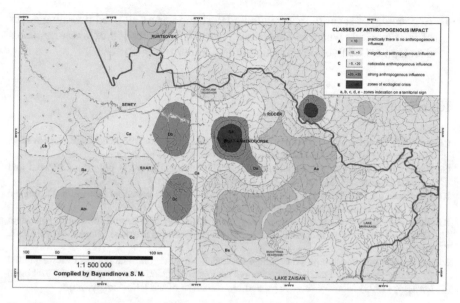

Fig. 3.2 Forecast zoning map of anthropogenous impact on the East Kazakhstan territory

From the forecast positions the zone of class E, ecological crisis in connection with development of mineral deposits are shifted to the south (Fig. 3.2).

Class A practically has lack of anthropogenous influence, and covers the smaller territory on comparison with previous classes, and has two centers of distribution named as Turgusun-Bukon (blocks 6–11, 6–12, 6–13, 7–12, 7–13,8–12, 9–9, 9–10), Karaulazek (8–2, 8–3, 8–4,9–2, 9–3,9–4). This zone includes special protected natural areas: national nature wildlife areas: Lower-Turgusun, Kulundzhun. Zone of insignificant and noticeable anthropogenous influence, includes the Markakol national nature park, Katon-Karagai national nature park.

Main conclusions on use in target function for division of anthropogenous impact on geoecosystems of East Kazakhstan, which has expanded set of parameters can be considered as the following.

1. The differentiated accounting of role and orientation concerning anthropogenous impact factors and natural factors in target function allowed to reveal broader set of classes on extent of anthropogenous impact on the area territory.
2. The highest level of anthropogenous impact (E zone classes) is the share of zones concerning industrial hubs influence with increased density of mineral deposits and manufacture industry. These zones demand priority actions on environmental protection.
3. Strong extent of anthropogenic impact (D class) is allocated on sites of industrial hubs, polymetallic fields, arable lands and pastures and also in zones of joint influence concerning several anthropogenic factors.

4. Accounting of security factor with underground waters allowed revealing zones of the first class, where in the actions presence for protection of water resources development of livestock production is optimum.

References

Chizhov AB, Gavrilov AV, Pizhankova EI (1995) K metodike geoekologicheskogo kartirovaniya (To technique of geoecological mapping) № 5. Geoecology, pp 88–95

Gerasimov IP (1986) Globalnyie i regionalnyie obschegeograficheskie prognozyi (Global and regional all-geographical forecasts). Vysshaya shkola, Moscow, pp 21–32

Konstantinov VM, Chelidze YB (2001) Ekologicheskie osnovyi prirodopolzovaniya (Ecological bases of environmental management). Academy, Moscow, p 714

Krauklis AA (1986) Geograficheskiy prognoz i rezultatyi izucheniya dinamiki geosistem (Geographical forecast and results of study on geosystem dynamics). Nauka, Novosibirsk, pp 95–117

Metodicheskoe rukovodstvo po provedeniyu ekologicheskogo rayonirovaniya territorii Respubliki Kazahstan (Methodical guide to carrying out ecological division of the Republic of Kazakhstan territory) (1995) Ministry of ecology and bio resources of the RK, Almaty, p 68

Chapter 4
Geoecological Bases of Nature Protection Measures and Actions

4.1 Problems of Nature Protection Activities in East Kazakhstan

Scientific organization principle of environmental management, which includes the uniform system of activities on ecological-economic character and social development of certain territory combining rational use of natural resources with optimum development and functioning of biogeosystems can be considered as basis for environmental management in a certain territory (Artusmanov 2001; Anuchin 1978).

Main aim is creation of the closed cycle economy based on five main activities (reduction, replacement, recovery, recirculation and reuse), ecologically rational circulation of materials, savings and replacements of non-renewable resources. The system of economic and organizational-legal conditions uniting and stimulating the solution of environmental protection problems on the basis of economic incentives has to be created for this purpose (Bazilevich 1987; konstantinov and Chelidze 2001).

As it is specified in NAPEP, on the basis of international experience, it is established that sustainable development of state economy is possible only at respect for the following basic principles:

- "inheritance of the benefits", present generation has to keep the biosphere for future generations;
- integrated approach to environment problems and society development;
  need of global cooperation for the benefit of ensuring sustainable development.

Environmental problems' solution of East Kazakhstan has to be based on integrated scientific approach. Rational use, preservation and restoration of land resources have to be the foundation as for economic and biological basis of human being existence. For the region with developed and long ago existing raw (including mining-metallurgical and power) complexes prime are problems of perspective transition to environmentally friendly, waste-free and low-waste technologies that will create a basis for stage-by-stage decrease in the current pollution levels. At the

© Springer Nature Singapore Pte Ltd. 2018
S. Bayandinova et al., *Man-Made Ecology of East Kazakhstan*,
Environmental Science and Engineering, https://doi.org/10.1007/978-981-10-6346-6_4

same time, these technologies have to resolve issues of processing, use or burial of already saved up industrial and household wastes that will allow resolving issues of stage-by-stage decrease in the existing level of industrial pollution. And, the introduced technologies should not resolve issues of quality improvement concerning one environment at the expense of others. In a complex, it will allow to provide stage-by-stage decrease in level of ecological impact on technogenic character concerning regions geosystems. During usage of water resource problems of preservation on their quality and quantity have to be solved. If the first task is connected with use by human being in the technological and household purposes, then the second depends on complex, both global reasons, and local, also first of all, on territories desertification. Thus, in East Kazakhstan preservation and restoration of the woods, forest shelter belts and corresponding biotas are prime for the second problem of water resources preservation. Pollutants distribution on the greatest distance comes from a source of pollution in the atmosphere. Transferred by air and water of substance from enterprises, the technogenic mineral educations (TME) including radioactive, municipal solid waste (MSW), from Semipalatinsk nuclear test site (SNTS) are deposited in soils, water deposits, biotas, and etc., becoming permanent negative factors on region ecology. All this complex of influences, and also overconsumption of natural resources, especially bioresources, define changes in geosystems, and influence biodiversity of territories, leading to their simplification and degradation. The possibility of bioresources wide use exists only at preservation and development of natural and unique geosystems under the state control (Owen 1977; Ohrana prirodyi 1987; Avessalomova 1992; Pentl 1979; Sokolov 1992).

Basis for analysis and decision-making has to be made in nature protection activity system of ecological monitoring. According to decision of the Government of RK, land registries of areas are issued as basis for creation of territorial ecological inventories, which are formed appropriately in an electronic form. In East Kazakhstan the land registry in required form (an electronic form) is not created yet, that is a first priority of ecological monitoring. Further development of ecological monitoring demands expansion of continuous data collection network from organized emission sources in the atmosphere and water, periodic territorial control in the territories are adjacent to industrial complexes. Besides, creation of periodic control system of water, land and bioresources is necessary. The collected data have to serve for creation and updating of the ecological inventory. Currently, the necessary collected minimum of data allowing to begin creation of the ecological inventory. Analysis and forecasting questions of an ecological situation development does not decide at all. This problem is closely connected with a question of qualified specialists training in the field of ecology.

4.2 Differentiation of Geosystems in East Kazakhstan for Development of Nature Protection Actions and Measures

On the basis of received map-schemes geosystem and current state of anthropogenous influence, as a result of imposing and differentiation, obtained map-scheme differentiation of geosystems for development of nature protection activities (Fig. 4.1, Table 4.1).

4.3 System of Nature Protection Actions for Geosystems in East Kazakhstan

Compensation of negative ecological factors at increase in the industry outputs demands holding the relevant activities. Main reason for the increase of ecological influence is exhaustion high degree of enterprises fixed assets. Clearing equipment operating on them was practically not upgraded within the last ten years. There was quite enough power at incomplete loading of the enterprises, but, considering its physical exhaustion, it is not capable to sustain increase in production any more. Many fields of mineral raw material parks were thrown and not taken out of service as appropriate. Updating coefficient dynamics on fixed assets of the industrial enterprises for the last three years shows the outlined tendency of its increase: in

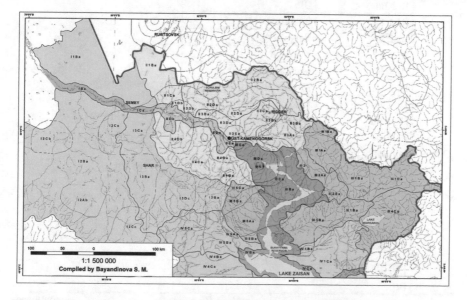

Fig. 4.1 Differentiation map of geosystems for development of nature protection activities

Table 4.1 Legend to the map-scheme differentiation of the East Kazakhstan geosystems for development of nature protection activities

Class	Index on map	Geosystems	System points of nature protection activities
1	2	3	4
Class A Practically there is no anthropogenous influence			
A	III Aa	Ertis-Bukhtyrma intrageosystem	III, IV, VII
	I 1 Aa	Shagan	
	II 3 Aa	Ulba	
	III 1 Aa	Bukhtyrma	
	III 2 Aa	Naryn	
	III 5 Aa	Koperly-Bukon	
	IV 4 Aa	Bugaz-Tebestin	
Class B Insignificant anthropogenous influence			
B	I Ba	Ertis intrageosystem	III, IV, VI, VII
	III Ba	Ertis-Bukhtyrma intrageosystem	
	IV Ba	Ertis-Zhaysan intrageosystem	
	I 2 Ba	Balapan-Ertis	
	I 2 Bв	Shagan	
	I 3 Ba	Shar	
	II 1 Ba	Schulbin	
	II 2 Ba	Uba	
	II 3 Ba	Ulba	
	III 1 Ba	Bukhtyrma	
	III 2 Ba	Naryn	
	III 3 Ba	Kurshym	
	III 4 Ba	Markakol-Karakubin	
	III 5 Ba	Koperly-Bukon	
	IV 1 Ba	Kalgaty-Takyr	
	IV 4 Ba	Bugaz-Tebestin	
	IV 5 Ba	Zhuzagash	
Class C Noticeable anthropogenous influence			
C	I Ca	Ertis intrageosystem	III, IV,VI VII
	II Ca	Ertis-Schulbin intrageosystem	
	III Ca	Ertis-Bukhtyrma intrageosystem	
	IV Ca	Ertis-Zhaysan intrageosystem	
	I 2 Cв, I 2 Ca, I 2 Cc	Shagan	

(continued)

Table 4.1   (continued)

Class	Index on map	Geosystems	System points of nature protection activities
	I 3 Ca	Shar	
	II 1 Ca	Schulbin	
	II 2 Ca, II 3 Ca	Uba	
	II 4 Ca	Kyzylsu-Taintin	
	III 1 Ca	Bukhtyrma	
	III 3 Ca	Kurshym	
	III 4 Ca	Markakol-Karakubin	
	III 5 Ca	Koperly-Bukon	
	IV 1 Ca	Kalgaty-Takyr	
	IV 4 Ca	Bugaz-Tebestin	
	IV 5 Ca	Zhuzagash	
Class D Strong anthropogenous influence			
D	II D$_B$, II Da	Ertis-Schulbin intrageosystem	I, II, V, VII
	III Da	Ertis-Bukhtyrma intrageosystem	
	II 1 D$_B$	Schulbin	
	II 2 Da	Uba	
	II 3 Da	Ulba	
	II 4 Da, II 4 D$_B$	Kyzylsu-Taintin	
Class E Zones of ecological crisis			
E	II Ea	Ertis-Schulbin intrageosystem	I, II, V, VII
	III Ea	Ertis-Bukhtyrma intrageosystem	
	II 3 Ea	Ulba	

1997—3,5%; in 1998—3,8%; in 1999—8,7%. In 1999 funds of the mining industry are updated on—21.7%, that is higher than previous year for 3, 5 times. In manufacturing industry—for 10.6% with growth for 1, 2 times; in production and distribution of the electric power, gas and water only for 3.6%. From the provided data, we can see that at exhaustion high degree of fixed asset rates on their updating lag behind requirements because of insufficient investments in all directions. But the basic problem solution of pollutants emission reduction is possible only at introduction and use of environmentally friendly technologies and, first of all, at the enterprises of mining-metallurgical complex.

Main categories of nature protection activities can be marked out as following:

1st-category of nature protection activities is connected with precautionary measures for protection of geosystem regional capabilities. Currently necessary conditions are introduction of special environmental assessment rules concerning any project or an economic action within the region.

2nd-category of nature protection activities is connected with direct measures for protection and security of aquatic and subaquatic geosystems, depending on geosystems condition. They consist of activities for fight against a silting, salinization, and bogging.

3rd-category of nature protection activities is connected with direct measures for protection of technogenic geosystems capability. This category is considered as the form, designed to keep shape of geosystems in a modern form without worsening it. Technogenic geosystems have the different features, necessary for protection and security of regional conditions, which are obligatory for preservation of this or that degree of functioning intensity. In order to realize this category of protection, there is a principle need of increase in technological level, its orientation to resource-saving and implementation of structural shifts in production of the region goods. Territorial distribution of such activities has to coincide with borders of technogenic geosystems.

4th-category of nature protection activities allow using of certain geosystems as resource, but on condition of rationalization on conclusion ways of their technogenic modifications from instability condition and restoration of nature—resource potential. Activities on accurately formulated requirements to reproduction of the mastered resource, first of all, concerning soil resources and vegetable complex belong to this category. The category includes preservation and restoration of soil natural connections vegetative-soil cover with protection options of separately considered water resources.

Major importance of this nature protection activities category is not to allow process of territorial dissociation on geosystems operating conditions of East Kazakhstan, which can lead to violation of its steady dynamic balance condition. Actually, within category resource and vulnerable groups of the phenomena and objects can be allocated. For the last, it is necessary to provide the special nature protection forms of government.

5th-category of nature protection activities combines natural-resource and production-ecological systems of activities. Here special measures for protection of water objects, soil and vegetable cover, recreational-improving complexes, landscapes allotted for settlement, which require the possibility of general prevention on negative processes connected with technogenic pollution of the region heavy metals. Principal value here gets a regulation of technological activities for conservation. Increase in use of nature protection cleaning constructions, organization and control of condition concerning technogenic environmental pollution. At the same time, effective measures for fight against technogenic pollution should be noted.

6th-category implements a complex of the prior nature protection activities designed to improve or return soil fertility, vegetable cover and rational use of water resources to an initial state. The category has production and ecological character.

7th-category of nature protection activities put problems on improvement of the economic mechanism concerning conservation regulation and rational environmental management, increase in natural-security education.

On the basis of obtained data analysis, geosystem differentiations of East Kazakhstan for nature protection activities (Fig. 4.1; Table 4.1), the marked-out categories according to the measures plan on realization of strategic tasks "Programs of rational use and protection of natural resources for 2001–2005", on implementation of the government program action of the Republic of Kazakhstan for 2000–2002 in the 4.6.2 "Creation of bases for the balanced use of natural resources" and the developed regional program of rational use and protection of natural resources, we offer system of nature protection activities for anthropogenous impactclasses for East Kazakhstan geosystems.

I. **On protection of atmospheric air for the E, D classes**:

- reduction of emissions total amount from stationary sources;
- reduction of emissions from non-stationary sources;
- improvement of technologies concerning mineral raw materials processing on mining-metallurgical complex enterprises;
- introduction of energy saving technologies on power system enterprises;
- at increase in the outputs, it is necessary to raise in equivalent volume efficiency of the gas treatment station equipment, by reconstruction of the operative equipment, input in a system of new implementations concerning certified systems of continuous control;
- to develop and introduce the effective gas treatment station equipment on large and small power system enterprises;
- carrying out activities on improvement of fuel combustion quality, use of absorber-catalysts, control toughening.

II. **On protection of water resources for the E, D classes**:

- additional facilities' construction for treatment facilities in the cities of Ust-Kamenogorsk, Semey;
- preservation of the thrown mineries and mines;
- improvement of sewage treatment technologies in region enterprises;

III. **On protection of land resources for the A, B, C classes**:

carrying out activities on the used lands restoration;
- introduction of soil-protective agriculture, restoration of correct crop rotation, state support of land users in respect of conducting ecologically balanced managing;
- creation of the state monitoring system for lands use.

IV. **On protection and reproduction of bioresources for the A, B, C classes**:

- improvement of the scheme development and placement of nature-park fund, and network of special protected nature areas that will allow to keep biodiversity, which is one of the factors on ecological equilibrium in region;

- strengthening of protection due to combination of all branches effort concerning the power of law enforcement and nature protection agencies of public organizations, and land user enterprises;
- creation of the uniform timber processing complex, having at the order is enough forces and means for certain forests development, introductions of deep wood processing and transition from use only of coniferous wood to deciduous;
- system recovery of forest conservation from the fires;
- protection of water biocenoses, preservation of hydrobionts biodiversity, gene pool of fishes rare species, optimization of bioresources use on area reservoirs.

V. **On utilization of industrial and household wastes for the E, D classes**:

- existing stores elimination of industrial wastes due to rational use;
- volumes reduction of dump product new receipts due to improved use and developments and deployments of new technologies on production and processing of mineral raw materials;
- processing and reduction of municipal solid waste intake, due to use of technology, ISO 14000 conforming to the international environmental standard;
- introduction of the international environmental standard of ISO 1400 on East Kazakhstan enterprises;
- introduction of forecast effective methods and technologies, searches and investigation of mineral deposits;
- to choose the optimum economic and ecologically acceptable project on utilization of municipal solid waste;

VI. **On radiational safety for the C classes**:

- to conduct comprehensive radiological examination of the territory;
- reduction of radionuclide content in natural water sources of the SNTS will resolve an issue of radionuclide conclusion from water recirculation of the SNTS industrial and technical complexes;
- to develop the concept of the SNTS economic land use;
- to resolve an issue of safe development on industrial and significant fields;
- creation of technology on solid radioactive waste utilization for the subsequent recultivation of the SNTS infected lands;
- to create system of radionuclides environmental monitoring;
- to resolve an issue of waste storage and processing on the industrial enterprises.

VII. **On regional environmental monitoring for the A, B, C, D, E classes**:

- creation of united regional information analytical center and development of the ecological inventory of the East Kazakhstan;
- using the provision on regional monitoring in the East Kazakhstan to create regional information center of environmental monitoring;
- to systematically carry out research works on definition of migration stability on environment pollutants;
- to develop models and methods of possible development forecasting on ecological situations in the region;

- expansion of minimum and sufficient network on environmental monitoring;
- to define minimum necessary, economically acceptable scheme of monitoring posts placement, taking into account character of controlled objects and their placement in the East Kazakhstan territory;
- to equip monitoring posts of the atmosphere, water resources, and radiation situation system on automatic control;
- to equip sources of stationary emissions by metrological certified systems of automatic control, for atmospheric emissions—systems of continuous control;
- to develop forecasting models and methods of ecological situations development in the region;
- databank creation of environmentally friendly technologies;
- to provide financial, personnel and technical security of bioresources and land resources monitoring.

Thus, creation of different categories on the nature protection activities, providing favorable conditions for human being existence, interests combination on different branches of the national economy as certain preservation complex of geosystems natural and resource capacity, in fact, it sets a task of geosystem—geochemical approach to implementation of the development concept and landscapes functioning at the regional level. During the realization of approaches, in determination of these or those categories on nature protection activities, expediently and accurately limit geosystems of the region, which need to be protected. Therefore, characteristics of its geosystem structure are important, and it is necessary for release of these or those ingredients, which are the main objects of protection in relation to geosystems.

Interaction complexity of geosystems, their regional features, problem of pollutants interenvironmental transition and development need on economy demands complex analysis of the economic growth reasons and consequences. It is possible only during creation of widely branched system on environmental monitoring and forecasting. Integrated approaches to environment issues need experts for the solution of ecological planning and rehabilitation problems. All decisions have to be carried out with broad support of the population and free access to ecologically significant information. For stable conditions creation of safe development in East Kazakhstan, the foundation of ecological education and education of younger generation has to be laid.

Solution of the existing environmental problems in the East Kazakhstan region, taking into account complex influence and interaction of environments and requirements of economy, perhaps can be implemented only on the basis of balanced scientific approach. The evidence-based regional program for rational use and protection of natural resources has to serve as the tool of environmental problems solution.

References

Anuchin VA (1978) Osnovyi prirodopolzovaniya. Teoreticheskiy aspekt (Environmental management bases). Theoretical aspect. Mysl', Moscow, pp 14–32

Artusmanov EA (2001) Ekologicheskie osnovyi prirodopolzovaniya (Ecological bases of environmental management). House of Dashkov and comp., Moscow, p 129

Avessalomova IA (1992) Ekologicheskaya otsenka landshaftov (Ecological assessment of landscapes). MSU, Almaty, p 88

Bazilevich NI (1987) Geograficheskie osnovyi ratsionalnogo prirodopolzovaniya (Geographical bases of rational environmental management). Nauka, Moscow, p 367

Konstantinov VM, Chelidze YB (2001) Ekologicheskie osnovyi prirodopolzovaniya (Ecological bases of environmental management). Academy, Moscow, p 714

Metodicheskoe rukovodstvo po provedeniyu ekologicheskogo rayonirovaniya territorii Respubliki Kazahstan (Methodical guide to carrying out ecological division of the Republic of Kazakhstan territory) (1995) Ministry of ecology and bio resources of the RK, Almaty, p 68

Ohrana prirodyi (Conservation) (1987) Reference book. Vysshaya shkola, Moscow, p 544

Owen OS (1977) Ohrana prirodnyih resursov (Protection of natural resources). Kolos, Moscow, p 681

Pentl R (1979) Metodyi sistemnogo analiza okruzhayuschey sredyi (Methods of the environment system analysis). Mir, Almaty, p 215

Sokolov VE (ed) (1992) Ekoinformatika: Teoriya. Praktika. Metodyi i sistemyi (Eco informatics: Theory. Practice. Methods and systems). Hydro Meteorological Publication, St. Petersburg, p 495

Chapter 5
Conclusion

The questions solution on stabilization and improvement of geosystems ecological condition of the East Kazakhstan, in conditions of intensive anthropogenous influence, demands the maximum accounting of concrete natural situation and the analysis of major factors and processes, which are negatively affecting its state. But in research region, the technogenic pollution is major destabilizing factor of geosystems functioning. It is important to note, what makes negative impact on condition of region geosystems, besides technogenic pollution by heavy metals and other pollutant, also anthropogenous change of the hydrological mode, water balance, hydrochemical indicators, reduction of geosystems soil fertility and their biological efficiency.

1. It is established, that territory of the East Kazakhstan represents the uniform megageosystem created by the Ertis river basin and consists of the interconnected and interacting elements of urbanized environment.
2. Role of geoecological study on physiographic factors defining the current state of geosystems is introduced: geographical location, Paleozoic denuded base, seven-arid climatic conditions, existence of dense river network, local features of soils and biota, intrazonal nature of valley geosystems. Certain role in destabilization of geosystems belongs to cross-border transfer of pollutants in a superficial drain.
3. Carried out analysis of the geosystems current state on nature of technogenic pollution, in the research region it is allowed revealing an overall picture of predicted geochemical situation, to estimate degree of technogenic loading with identification of existing spatial manifestation and estimated risk of geosystems technogenic destabilization with tension zones definition.
4. It is established, that the current state of the East Kazakhstan geosystems is characterized by intense level of environment destabilization, which is shown in essential, often irreversible changes of hydroclimatic factors and geosystems biotic structure.

© Springer Nature Singapore Pte Ltd. 2018   135
S. Bayandinova et al., *Man-Made Ecology of East Kazakhstan*,
Environmental Science and Engineering, https://doi.org/10.1007/978-981-10-6346-6_5

5. Differentiated accounting of factors role and orientation on anthropogenous influence and natural factors in target function allowed revealing broader set of classes on extent of anthropogenous impact in the area territory.
6. The highest level of anthropogenous influence (E class zone) is share of industrial hubs influence zones with increased density of mineral deposits and processing enterprises. These zones demand priority actions on environmental protection.
7. Strong extent of anthropogenous influence (class D) is allocated on sites of industrial hubs, polymetallic fields, arable lands and pastures, and also in joint influence zones of several anthropogenous factors.
8. Regional nature protection activities for security of the East Kazakhstan environment are developed on the basis of differentiation and zone of geosystems anthropogenous loading in the East Kazakhstan. They allow defining the concrete directions of geosystems optimization in region.
9. Research results, which are received above, can be used at further detailed research of geosystem dynamics on functioning under the technogenesis influence. Cartographic materials and offered nature protection activities will allow developing optimal variants of the problems solution concerning natural resources complex use, and also can be used by the production, scientific and other organizations, setting purpose in the problems solution of environmental protection and rational environmental management.

Further improvement and development of theoretical, methodical and applied provisions stated in the thesis, has to be directed to questions of basic improvement on natural resources use quality, for providing fuller solution of practical and economic tasks in East Kazakhstan geosystems.

Appendix A
Legend to Landscape Maps of the East Kazakhstan

Class of Flat Landscapes

Dry Steppe Landscapes

Accumulative lacustrine—alluvial, semihydromorphic

1. Hilly-elevated plain with temporary water temporary courses, put by sandy loams and loams with forb-fescue feather grass vegetation with fragments of the pine woods on chestnut soils.
2. Hilly-elevated plain with small lakes and decreases put by loams with fragments of the pine woods and forb-fescue feather grass communities on meadow with meadow solonetzic soils.
3. Sublime-undulating plain—wavy plain with ravine network put by loams, sandy loams, with forb-fescue feather grass vegetation on chestnut normal soils and the pine wood fragments on sands.
4. Sublime-undulating plain—wavy plain with kettle and small hollow hills put by loams and sandy loams with forb-porcupine-fescue feather grass vegetation on dark-chestnut soils.
5. Hilly plain—flat plain with sai put by sandy loams, loams, sands with fescue-feather grass, fescue-goldilocks and camphoric vegetation, with inundated willow thickets, on steppe and chestnut solonetzic soils.
6. Hilly plain flat terraced hollow with kettle put by sands and clays with fescue feather grass, fescue-wormwood, fescue-goldilocks, wormwood and camphoric vegetation on meadow solonetzic soils.

Denudation, Automorphous

7. Elevated hilly plain put by effusive-sedimentary breeds with Austrian-wormwood-tyrsovo-fescue and fescue-feather grass steppe and pea shrub

© Springer Nature Singapore Pte Ltd. 2018
S. Bayandinova et al., *Man-Made Ecology of East Kazakhstan*,
Environmental Science and Engineering, https://doi.org/10.1007/978-981-10-6346-6

vegetation with participation of reedgrass-chi-fescue meadows on chestnut normal soils with steppe solonetzic soils.

8 Sublime-hilly steppe plain put by effusive-sedimentary breeds with shrubby-wormwood and feather grass vegetation on chestnut undeveloped soils with solonetz.

3. Sloping weak hilly steppe plain put by sedimentary-effusive breeds with shrubby-fescue-oat grass vegetation on dark-chestnut solonetzic soils.

4. Hilly plain with small bald peak put by sandstones, tuffaceous sandstone, granites with shrubby-fescue-oat grass vegetation on dark-chestnut solonetzic soils.

5. Dome-hilly plain with bald peak put by granites, granodiorite with shrubby-cereal-cold wormwood vegetation on dark-chestnut soils.

Semi-deserted Landscapes

Accumulative lacustrine—alluvial (neo-eluvial, paleohydromorphic)

12. Gently-sloping plain put by loams, clays, sands with shrubby-cereal-cold wormwood vegetation on light brown normal solonetzic soils.

13. Gently-sloping plain put by loams, clays, gravel with wormwood-fescue vegetation and participation of shrubby-fescue-feather grass associations on light brown solonetzic soils.

14. Gently-sloping plain put by loams, clays, gravel with shrubby-wormwood and feather grass vegetation on light brown normal solonetzic soils.

Denudation, Automorphous

15. Gently-sloping plain put by effusive-sedimentary breeds with forb-fescue feather grass and fescue-feather grass steppe and pea shrub vegetation on the light brown not full-developed soils with steppe solonetz.

16. Hilly-steeply sloping plain with hills put by effusive-sedimentary breeds with shrubby-wormwood and feather grass vegetation on light brown solonetzic carbonate soils.

17. Dome-hilly-ridge plain put by effusive-sedimentary breeds with shrubby-wormwood vegetation on the light brown not fully-developed and normal soils with solonetz.

Desert Landscapes

Accumulative Neo-eluvial

18. Hilly-wavy plain put by loams, sandy loams, sands with grey wormwood, ephemeral-grey wormwood, dormouse-grey wormwood vegetation on meadow-brown soils with solonetz and saline soils.
19. Dome-shallow ridge plain put by effusive-sedimentary breeds with cereal-wormwood vegetation on brown soils, normal with solonetzic soils.

Class of Mountain Landscapes

Nival Landscapes

20. Highlands with the Alpine relief forms, modern freezing put by granites, granodiorite, gabbro with single flowering plants, lichens, mosses on primitive soils.

Meadow Landscapes, Paleohydromorphic

21. Middle mountains of alignment surfaces put by sandstones, quartzites, limestones, granites with subalpine and Alpine meadows on mountain meadow soils.
22. Highlands with left surface alignment, and old glacial period forms of relief, put by slates, quartzites, gneisses, jaspers, subalpine and Alpine meadows, meadow-steppes, meadow-saz on mountain meadow and mountain meadow-steppe soils.

Forest Landscapes

Low Mountain Landscapes, Automorphous

23. Hilly-steeply sloping middle mountains put by granites, with the pine woods on mountain chestnut soils.
24. Steeply sloping low mountains, put by effusive-sedimentary thickness, with the fir wood and meadow sites on mountain-forest (sour) nonpodzolized and mountain-meadow soils.

25. Gravelly low mountains, put by porphyrite, tuffs, tuffaceous sandstone with birch-aspen woods and meadow steppes on mountain black earth, forest and meadow-steppe soils.

Middle Mountain Landscapes

Neo-eluvial

26. Ridge middle mountains put by sedimentary-effusive thickness with mountain larchen woods and bushes on mountain black earth and mountain-forest gray soils.
27. Ridge middle mountains put by limestones, aleurolites, gravelstone, sandstones with the fir woods, meadow sites on mountain-forest (podzolized sour) and mountain-meadow soils.
28. Ridge middle mountains put by limestones, aleurolites, gravelstone, sandstones with larchen woods on mountain-forest gray nonpodzolized soils.
29. Ridge middle mountains put by sedimentary-effusive breeds with the fir woods on mountain-forest gray podzolized soils.

Steppe Landscapes

Foothill Landscapes, Transeluvial

30. Hilly-foothill plain put by loams, gravel-rubbish material, with wormwood-tyrsovy vegetation on light brown normal soils.
31. Hilly-foothill plain put by loams, gravel-rubbish material with vegetation on chestnut soils.
32. Hilly-steeply sloping foothill plain put by effusive-sedimentary breeds with shrubby-fescue-feather grass vegetation with birch splittings, on black earth of ordinary leached.
33. Steeply sloping-ridge foothill plain put by granites, granodiorite with shrubby oat grass-feather grass vegetation on black earth ordinary leached and podzolized.
34. Steeply sloping plateau put by effusive-sedimentary breeds with shrubby-fescue-feather grass vegetation on black earth of the southern normal.
35. Steeply sloping plateau put by effusive-sedimentary breeds with shrubby-oat grass-feather grass vegetation on black earth of the southern normal.
36. Hillside plateau put by effusive-sedimentary breeds with shrubby-fescue-feather grass vegetation on mountain chestnut soils.

37. Hillside plateau put by effusive-sedimentary breeds with shrubby-oat grass-feather grass vegetation on mountain-chestnut soils.
38. Hillside plateau put by sandstones, tuffs with shrubby-wormwood vegetation on chestnut and mountain-chestnut soils.
39. Hillside plateau put by sandstones, tuffs with shrubby-fescue-feather grass vegetation on ordinary black earth.
40. Hillside plateau put by sedimentary-effusive deposits with shrubby-wormwood-kovylkovy vegetation on chestnut undeveloped soils in combination with solonetzic soils.
41. Hillside plateau put by forested loams, boulder-gravel, and shrubby-wormwood vegetation on dark-chestnut soils.
42. Hillside plateau put by porphyrite, tuffs, sandstones with shrubby-wormwood vegetation on dark-chestnut soils.
43. Hillside plateau put by porphyrite, tuffs, sandstones with shrubby-cold wormwood-cereal vegetation on mountain chestnut soils.

Low Mountain Landscapes

Automorphous

44. Ridge-steeply sloping put by sedimentary-volcanogenic breeds, granites with fescue-feather grass mountain vegetation, with participation of bushes, forb meadows on mountain chestnut soils.
45. Ridge-steeply sloping put by sedimentary-volcanogenic breeds, granites with shrubby-cereal vegetation on mountain chestnut soils.
46. Hilly-steeply sloping put by sandstones, tuffaceous sandstone, tuffaceous pro-phyrite with shrubby-oat grass-feather grass vegetation and shrubby-fescue-feather grass vegetation on mountain black earth and mountain-chestnut soils.
47. Hilly-steeply sloping put by sandstones, tuffaceous sandstone, tuffaceous pro-phyrite with shrubby-fescue-feather grass vegetation on ordinary black earth.
48. Hilly-steeply sloping put by sandstones, tuffaceous sandstone, tuffaceous pro-phyrite with large cereal forb vegetation and bushes on mountain chestnut soils.
49. Hilly-steeply sloping put by sandstones, tuffaceous sandstone, tuffaceous pro-phyrite with shrubby-cold wormwood-cereal vegetation on mountain chestnut and mountain-steppe the xeromorphic soils.
50. Wavy-hilly put by slates, gneisses, sandstones, with Austrian-wormwood-tyrsovo-feather grass vegetation on mountain dark-chestnut soils.
51. Wavy-hilly put by granites, deprived of vegetable cover, on valleys birch-aspen-willow shrub scaffold with shrubby-wormwood vegetation on mountain chestnut soils.
52. Wavy-hilly, with left alignment surfaces, put by sandstones, slates, limestones, granites, shrubby-fescue-feather grass vegetation on mountain chestnut soils.

Middle Mountain Landscapes

Automorphous

53. Folded-horst put by prophyrite, tuffs, sandstones, with fescue-feather grass mountain vegetation with participation of bushes, forb meadows on mountain black earth and mountain chestnut soils.

Landscapes of Intra Mountain and Intermountain Hollows

Semihydromorphic

54. Flat-sloping plain put by loams, bouldery-gravel, with wormwood fescue-vegetation and forb-cereal meadows on mountain chestnut and meadow soils.
55. Flat-sloping plain put by loams, bouldery-gravel-channery material, fescue-forb vegetation with participation of bushes on mountain chestnut soils.
56. Hilly-wavy plain put by loams, sands, clays with forb-cereal vegetation on meadow soils with solonetz.

Semideserted Landscapes

Foothill Landscapes, Transeluvial

57. Hilly plain put by loams, bouldery-gravel, with shrubby-wormwood vegetation on chestnut and mountain chestnut soils.
58. Hilly plain put by loams, bouldery-gravel with wormwood vegetation on light brown carbonate soils.
59. Steeply sloping plain put by clays, sands with wormwood vegetation and participation of bushes on light brown soils.
60. Steeply sloping-hilly plain put by sandstones, tuffaceous sandstone, effusive, tuffs with wormwood vegetation on brown normal and light brown undeveloped soils with participation of solonetzic soils, and light brown solonetzic soils.
61. Steeply sloping-hilly plain put by sedimentary-effusive thickness with lessingit-wormwood and cereal-lessingit-wormwood vegetation on chestnut soils.

Low Mountain Landscapes

Automorphous

62. Low mountain put by sandstones, conglomerates, effusive, tuffs, with shrubby-wormwood-fescue vegetation on the mountain chestnut not full-developed soils.

Landscapes of Intra Mountain and Intermountain Hollows

Transeluvial

63. Ridge-hummocky plain put by sands with forb-erkekovy, wormwood-erkekovy vegetation on sands and gray-brown soils.
64. Ridge-hummocky plain put by sands with forb-wormwood-erkekovy vegetation and participation of pine on gray-brown soils and sands.

Desert Landscapes

Foothill Landscapes, Transeluvial

65. Wavy-sloping plain put by forested loams, bouldery-gravel with ephemeral-dormouse-grey wormwood vegetation on gray-brown normal soils.
66. Wavy-sloping plain put by forested loams, bouldery-gravel from cereal-grey wormwood vegetation, chi, azhrekovy meadows on sierozemic and meadow-sierozemic solonetzic soils.

Landscapes of Intra Mountain and Intermountain Hollows

Semihydromorphic

67. Concave plain with hollows put by loams, sandy loams, clays, with sarzasan, karabaran, obion vegetation on gray-brown normal, meadow-brown and meadow-sierozemic soils with meadow solonetzic soils.
68. Concave-hilly plain with small hills and ephemeral, eurotia, grey wormwood and calligonum groups on sands in combination with solonetzic soils and takyr-like soils.

69. Weak hilly steppe plain with courses of temporary water currents put by loams, sandy loams, sands, clays, with grey wormwood, ephemeral-grey wormwood, keyreukovo-grey wormwood vegetation on gray-brown soils with solonetzic soils.
70. Flat-concave plain with small sandy manes put by oozy clays, loams, sands with reed and clubroot vegetation on meadow and marsh saline soils.

Valley Landscapes

Hydromorphic

71. Flood plains put by loams, sandy loams, sands, gravel with forb-cereal meadows and small-leaved woods on nominal-meadow soils.
72. Flood plains put by clays, loams, sands with forb-cereal meadows, willow thickets on alluvial soils.
73. Flood plains put by clays, loams with quackgrass, bromegrass, sedge saline meadows on meadow solonetzic-saline soils and meadow solonetzic soils.
74. Flood plains put by loams, sandy loams with chi, wild rye, azhrekovy meadows on alluvial soils.
75. Flood plains put by loams, sands with riparian forests on alluvial and meadow-soils.

Printed in the United States
By Bookmasters